ADVANCED RESERVOIR AND PRODUCTION ENGINEERING FOR COAL BED METHANE

ADVANCED RESERVOIR AND PRODUCTION ENGINEERING FOR COAL BED METHANE

PRAMOD THAKUR
ESMS LLC, Morgantown, WV, USA

ELSEVIER

AMSTERDAM • BOSTON • HEIDELBERG • LONDON
NEW YORK • OXFORD • PARIS • SAN DIEGO
SAN FRANCISCO • SINGAPORE • SYDNEY • TOKYO
Gulf Professional Publishing is an imprint of Elsevier

G|P
P|

Gulf Professional Publishing is an imprint of Elsevier
50 Hampshire Street, 5th Floor, Cambridge, MA 02139, United States
The Boulevard, Langford Lane, Kidlington, Oxford, OX5 1GB, United Kingdom

British Library Cataloguing-in-Publication Data
A catalogue record for this book is available from the British Library

Library of Congress Cataloging-in-Publication Data
A catalog record for this book is available from the Library of Congress

ISBN: 978-0-12-803095-0

Publisher: Joe Hayton
Senior Acquisition Editor: Katie Hammon
Senior Editorial Project Manager: Kattie Washington
Production Project Manager: Kiruthika Govindaraju
Cover Designer: Maria Ines Cruz

Typeset by MPS Limited, Chennai, India

DEDICATION

The book is dedicated to the safety of all coal miners and gas production from coal.

CONTENTS

PREFACE

The current world human population of over 7000 million people consumes about 5×10^{17} BTU of energy per year. This is expected to increase to 7.5×10^{17} BTU per year by 2040. About 87% of all energy consumed comes from fossil fuels, and nuclear and hydropower provide 12%. Solar, wind and geothermal energy provide less than 1%. Among fossil fuels (oil, natural gas and coal), about 85% of the available energy is contained in coal although only 26% of all energy consumed is derived from coal. It is therefore very likely that coal's share of the energy mix will increase in the future. Global coal deposits are widespread in 70 countries. Coal is the most abundant and economical fuel today, costing only 4 cents/kWh of electricity. The mineable reserve of coal (to a depth of 3000 ft) is about 1 trillion tons, but the total indicated reserve to a depth of 10,000 feet is between 17 and 30 trillion tons. Exploitation of the energy contained in the nonmineable coal reserve is the essence of this book.

Besides coal, this vast coal reserve contains another source of energy, coalbed methane (CBM). It is almost like natural gas with about 10–15% lower calorific value. Reserve estimates of CBM ranges from 275 to 34,000 TCF. This huge reserve of gas remains almost unexploited. CBM production only started in the 1980s and the current global production is 3 TCF/year. About 60% of this production is in the United States. Coal and CBM are syngenetic in origin; thus, coal is both the source and the reservoir for CBM. Coal seams are formed over millions of years by the biochemical decay of plant materials. The process produces vast amount of methane and carbon dioxide as the plant materials metamorphose to coal. Most of the gas escapes to the atmosphere and only a small fraction is retained in coal. The gas content of coal ranges from 35 to 875 ft^3/ton to a depth of 4000 ft. Not much data is available for deeper coal seams but, in general, the gas content increases with depth.

The coal reservoir for gas significantly differs from conventional gas reservoirs, requiring a separate treatment of the subject. The book discusses all aspects of reservoir engineering and production engineering for CBM.

The material for the book was developed by the author to teach a graduate-level course on the subject at the West Virginia University over the past 10 years. It takes a 15-week-long semester to cover this course. The content is based on the author's derivation of mathematical equations for measuring reservoir properties and his forty years of experience in the field on production engineering. This work is strictly a graduate level book but can be later expanded to include undergraduate material to make it amenable for two courses: an undergraduate and a graduate-level course.

Chapter 1, Global Reserves of Coal Bed Methane and Prominent Coal Basins, is a general introduction to global CBM reservoirs with a description of current gas production activities. Nineteen basins are identified, and the geology and reserves of prominent basins are discussed.

The next four chapters (chapters Gas Content of Coal and Reserve Estimates, Porosity and Permeability of Coal, Diffusion of Gases From Coal, and Pore Pressure and Stress Field in Coal Reservoirs) present the reservoir engineering aspects of CBM. Chapter 2, Gas Content of Coal and Reserve Estimates, discusses gas contents and gas isotherms of coal seams with methods for the estimation of gas-in-place (GIP) for both mineable and non-mineable reserves. Chapter 3, Porosity and Permeability of Coal, deals with the porosity and permeability of coal seams. Definitions of terms and various methods of measuring or estimating permeability are discussed. Chapter 4, Diffusion of Gases From Coal, derives equations for the measurement of diffusivity and sorption time. Chapter 5, Pore Pressure and Stress Field in Coal Reservoirs, deals with reservoir (pore) pressure and ground stress. The influence of these stresses on production technology is also discussed.

Chapter 6, Fluid Flow in Coal Reservoirs and Chapter 7, Fluid Flow in Pipes and Boreholes discuss the flow of fluids in porous media, such as coal and shale, and the flow of fluid in pipes and gas wells. They provide mathematical equations to calculate gas production and reservoir pressure decline, which are essential for efficient gas production. The flow of slurries in pipes and pipe annulus are also discussed. Gas production decline is discussed in detail.

Chapter 8, Hydraulic Fracking of Coal, deals with hydrofracking of coal. This is currently the most popular method of gas production. Hydrofracking of both vertical and horizontal wells is discussed. Sand schedules for water and nitrogen-foam fracking are provided. One of the unique contents of this chapter is the in-mine measurements of the length, width, and height of the fracture and verification of extant theories. Some 200 wells were mapped and results are summarized.

Chapter 9, Horizontal Drilling in Coal Seams, deals with horizontal drilling both in-mine and from the surface. This is the future technique for CBM production from deep coal seams. Design of drill rigs for in-mine use is discussed.

Finally, twelve US coal basins are classified on the basis of depth as (a) shallow, (b) medium-depth, and (c) deep basins. The most suitable production technique for each basin is presented in addition to a summary of current CBM production activities. Secondary recovery of CBM by CO_2 flooding and tertiary recovery of energy in coal by underground coal gasification are very briefly discussed.

Most of the knowledge contained in the book was discovered by the author while working for CONOCO/CONSOL Energy (an erstwhile subsidiary of CONOCO) from 1974 through 2014.

I gratefully acknowledge the help and guidance provided to me on hydrofracking of coal by my two friends, the late Dr. H.R. Crawford of CONOCO and the late Fred Skidmore of Texas. I am grateful to the late Eustace Frederick for his support in using these techniques to degas the world's gassiest mine, which produced 70 MMCFD of methane.

For the development of in-mine drilling rigs, I owe thanks to the late William Poundstone of CONSOL Energy, who mentored me, and the late Robert Fletcher of JH Fletcher Co, for manufacturing the first in-mine drilling rig for the coal industry.

I would be remiss in my duty if I did not thank Joyce Conn, who has typed most of my publications for the last 42 years, and Kattie Washington of Elsevier for her patience and guidance.

This book can be considerably expanded in the second edition if the need arises.

CHAPTER 1

Global Reserves of Coal Bed Methane and Prominent Coal Basins

Contents

Abstract

Fossil fuels comprise nearly 90% of the proved reserves of global energy. Coal is the major component of fossil fuel containing nearly 90% of the fossil fuel energy. The growing population of the world would need 5 to 7.5 × 10^{20} J of energy to live well. To meet this growing demand extraction of gas contained in coal has become necessary. The vast deposits of coal (17–30 T tons) contain approximately 30,000 TCF of gas, called coal bed methane (CBM). A brief description of the prominent coal basins with a CBM reserve estimate is provided. The list includes coal basins of United States, Western Canada, United Kingdom, France, Germany, Poland, Czech Republic, Ukraine, Russia, China, Australia, India, and South Africa. These countries produce 90% of global coal production and nearly 100% of all CBM production. Since the economic depth limit for mining is around 3000 ft, only about 1 T ton of coal can be mined leaving a vast reserve of coal full of CBM unutilized. Vertical drilling with hydrofracking (a copy of conventional oil and gas production technique) is the main technique used to extract gas at present. This works only up to 3000–3500 ft depth because of serious loss in permeability. A new technique that has been eminently successful in deep and tight Devonian Shale (Marcellus Shale) is advocated for CBM production from deeper horizons. Lastly the CBM reservoir is compared to conventional reservoirs.

Advanced Reservoir and Production Engineering for Coal Bed Methane.
DOI: http://dx.doi.org/10.1016/B978-0-12-803095-0.00001-6

The disparities are very substantial warranting a new and proper treatment of the subject, "Reservoir and Production Engineering of Coal Bed Methane."

1.1 INTRODUCTION

Coal seams were formed over millions of years (50–300 million) by the biochemical decay and metamorphic transformation of the original plant material. The process known as "coalification" produces large quantities of by-product gases. The volume of by-product gases (methane and carbon dioxide) increases with the rank of coal and is the highest for anthracite at about 27,000 ft^3/t (765 m^3/t) of methane [1]. Most of these gases escape to the atmosphere during the coalification process, but a small fraction is retained in coal. The amount of gas retained in the coal depends on a number of factors, such as the rank of coal, the depth of burial, the immediate roof and floor, geological anomalies, tectonic forces, and the temperature prevailing during the coalification process. In general, the higher the rank of coal and the greater the depth of coal, the higher is the coal's gas content. Actual gas contents of various coal seams to mineable depths of 4000 ft (1200 m) indicate a range of 35–875 ft^3/t (1–25 m^3/t).

Methane is the major component of coal bed methane (CBM), accounting for 80–95%. The balance is made up of ethane, propane, butane, carbon dioxide, hydrogen, oxygen, and argon. Coal seams are, therefore, both the source and reservoir for CBM.

1.2 COAL AND CBM RESERVES

Coal is the most abundant and economical fossil fuel resource in the world today. Over the past 200 years, it has played a vital role in the stability and growth of the world economy. The current world human population of about 7000 million consumes 5×10^{20} J of energy per year. It is expected to increase to 7.5×10^{20} J/year in the next 20 years. About 87% of all energy consumed today is provided by fossil fuels. Nuclear and hydropower provide 12%. Solar, wind and geothermal energy provide barely 1% [2] as shown in Table 1.1.

Barring a breakthrough in nuclear fusion, fossil fuels will remain the main source of energy in the foreseeable future, as they have been for the past 200 years. Ninety percent of all fossil fuel energy in the world is in coal seams. It is, therefore, essential that coal's share in the energy mix should increase. At present, coal provides 26% of global energy demand

Table 1.1 World energy reserves & consumption

Fuel type	Energy consumed (EJ/y)[a]	Proved reserves (ZJ)[b]
Coal	120	290
Gas	110	15.7
Oil	180	18.4
Nuclear	30	2–17[c]
Hydro	30	N.A.
All others	4	Uncertain

[a]E = 10^{18}.
[b]Z = 10^{21}.
[c]Reprocessing not considered. 1000 J = 0.948 BTU.

Table 1.2 Global coal production

Country	Annual production (metric)[a] t (2013)
China	3561
United States	904
India	613
Indonesia	489
Australia	459
Russia	347
South Africa	256
Germany	191
Poland	143
Kazakhstan	120

[a]1 metric ton = 1.1 short tons.
Source: Adapted from World Coal Statistics. World coal association, <http://worldcoal.org/>; 2013, [4].

and generates 41% of the world's electricity. Coal deposits are widespread in 70 countries of the world. Coal is a very affordable and reliable source of energy. The total proved, mineable reserve of coal exceeds 1 T ton to a depth of about 3300 ft (1000 m). Indicated reserves (mostly nonmineable) to a depth of 10,000 ft (3000 m) range from 17 to 30 T ton [3]. Current (2014) world coal production is about 8000 million ton/year. Coal production from the top ten countries is shown in Table 1.2.

Total tonnage mined in these 10 countries comprise nearly 90% of global production. Coal production may continue to increase if they start converting coal into synthetic gas and liquid fuels, such as, diesel and aviation fuels.

Table 1.3 Estimates of CMB reserve

Country	[a]Estimated coal reserve (10^9 tons)	1992 estimated TCF (Tm^3)	1987 Estimated TCF (Tm^3)
United States	3000	388 (11)	30–41
Russia	5000	700–5860 (20–166)	720–790
China	4000	700–875 (20–25)	31
Canada	300	212–2682 (6–76)	92
Australia	200	282–494 (8–14)	N.A.
Germany	300	106 (3)	2.83
India	200	35 (1)	0.7
South Africa	100	35 (1)	N.A.
Poland	100	106 (3)	0.4–1.5
Other countries	200	177–353 (5–10)	N.A.
Total gas in place (GIP)		275–11,296 (78–320)	30,958–33,853 (877–959)

Recoverable Reserve: 30–60% of GIP.
US Conventional Gas Reserve [875] TCF ([25] Tm^3).
[a]US EPA, 2009 [7].

Besides the minable coal reserve, the vast deep-seated deposits of coal contain another source of energy; CBM. It is almost like natural gas with a slightly lower (10–15%) calorific value. Reserve estimates of CBM in coal ranges from 275 to 33,853 TCF (78–959 Tm^3) [5,6] as shown in Table 1.3.

Fig. 1.1 shows the major coal basins around the world [8]. A brief description of only reservoir data for prominent basins is provided below. Other information is provided in the book *Coal Bed Methane from Prospect to Pipeline* [9].

1.2.1 US Coal Basins

Not counting Alaska, there are three major basins in the United States, as shown in Fig. 1.2.

1. The Western United States
2. The Illinois basin
3. The Appalachian basin

These regions can be further divided into 14 sub-basins with additional information [10], but only larger basins are discussed here.

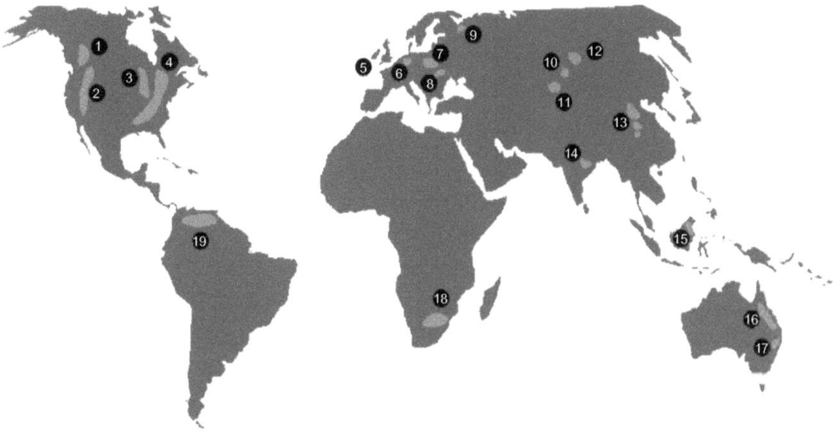

Figure 1.1 Major coal basins of the world.

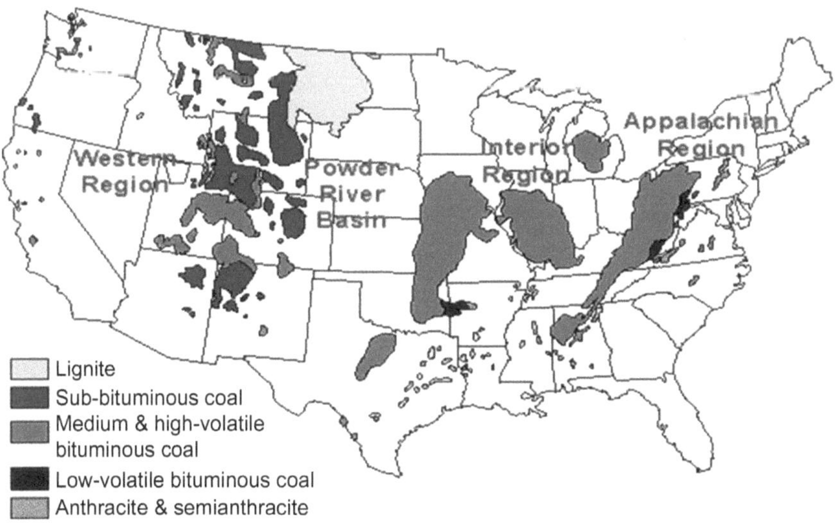

Figure 1.2 Major US coal basins.

1.2.1.1 The Western United States. It has the Following Sub-Basins

1. *San Juan basin* (Colorado, New Mexico). The basin covers an area of 14,000 mile^2. Coal seams with a thickness of 15–50 ft occur to a depth of 6500 ft with a total thickness of 110 ft. Gas content of coal varies from 300 to 600 ft^3/t. The coal is of low to high volatile bituminous rank. Permeability is medium to high (1–50 md). It is a well-developed field. Gas production is done by vertical, hydrofracked wells. Introduction of a new technology, horizontal boreholes (BH) drilled from the surface with hydrofracking, can increase gas production dramatically. A part of this basin is over-pressurized (gas pressure higher than hydrostatic pressure) leading to very high gas productions.
2. *Piceance basin* (W. Colorado). The basin covers an area of about 7000 mile^2. The thickness of coal seams varies from 20 to 30 ft with a total of up to 200 ft. The depth of coal seams varies from outcrop to 12,000 ft. The gas content of coal is 400–600 ft^3/t. The coal is of low grade. Permeability is generally low but there are areas where permeability of 1–5 md is indicated. Gas production is achieved by vertical drilling and hydrofracking. Gas production can be greatly increased by drilling horizontal BH from the surface and hydrofracking.
3. *Powder River basin* (Wyoming and Montana). This is a large basin covering 26,000 mile^2. The thickness of coal seams varies from 50 to 200 ft with a total coal thickness of 150–300 ft. The depth of coal seams varies from outcrop to 2500 ft. The gas content of coal is low at about 70 ft^3/t. The rank of coal is low: lignite to sub-bituminous. Permeability is usually high, ranging from 50 to 1500 md. Current gas production is from a shallow depth of 1000 ft or less. Gas production is achieved by vertical wells. Because of high permeability, no hydrofracking is generally needed.

1.2.1.2 The Illinois Basin (Illinois, Kentucky, and Indiana)

This is one of the largest basins, with an area of 53,000 mile^2. The thickness of coal seams varies from 5 to 15 ft with a total of 20–30 ft. The depth of cover varies from 0 to 3000 ft. The rank of coal ranges from HVC to HVB bituminous. The gas content is low, from 50 to 150 ft^3/t. The permeability is high in shallow areas approaching 50 md. Gas production is realized by vertical drilling and hydrofracking to a depth of generally less than 1000 ft. Hydrofracking of shallow coal is very inefficient. Horizontal BH drilled from the surface would be much more productive. No hydrofracking of horizontal BH is needed at shallow depths up to 1500 ft.

1.2.1.3 The Appalachian Basin (Pennsylvania, West Virginia, Virginia, Ohio, Maryland, Kentucky, Tennessee, and Alabama)

From a gas production point of view, the basin can be divided into two regions:

a. Northern Appalachia, and
b. Central and Southern Appalachia

1.2.1.3.1 Northern Appalachian Basin (Pennsylvania, West Virginia, Ohio, Kentucky, and Maryland)

This is a large basin with an area of about 45,000 mile^2 and it has been most extensively mined for the past 100 years. The thickness of coal seams vary from 4 to 12 ft with a total thickness of 25–30 ft. The depth of cover varies from 0 to 2000 ft. The coal rank varies from low volatile to high volatile bituminous coal. The gas content of coal seams ranges from 100 to 250 ft^3/t. The permeability to a depth of 1200 ft varies from 10 to 100 md. Gas production is mainly realized by drilling horizontal BH in the coal seam from the surface and in-mine workings. The specific gas production from various coal seams varies from 5 to 20 MCFD/100 ft of horizontal BH. This is the initial production from a freshly drilled BH. The total gas reserve is 61 TCF.

1.2.1.3.2 Central and Southern Appalachian Basin (West Virginia, Virginia, Kentucky, Tennessee, and Alabama)

The combined area of the basin is about 46,000 mile^2. The thickness of coal seams varies from 5 to 10 ft with a total thickness of 25–30 ft. The depth of cover varies from 1000 to 3000 ft. Mining depth generally does not exceed 2500 ft. The gas content of coal varies from 300 to 700 ft^3/t. The rank of coal is from low vol. to high vol. A bituminous. The permeability of coal seams varies from 1 to 30 md. Specific gas production from horizontal BH is 5–10 MCFD/100 ft. The main gas production technique is vertical drilling with hydrofracking. For commercial gas production, multiple coal seams are hydrofracked in a single well. Gas production of 250–500 MCFD is quite common for a single well completed in 3–5 coal seams. The total gas reserve is estimated at 25–30 TCF.

The CBM industry in the United States is well established. Nearly 50,000 wells have been completed with a total annual production of 1.8 TCF (about 10% of total US gas production). It can be easily doubled if the new technology of horizontal BH drilled from the surface and hydrofracking is applied to western thick coal seams.

1.2.2 Coal Basins of Canada

Most of the coal deposits in Canada are located in the provinces of Alberta and British Columbia, spread over a vast area of almost 400,000 $mile^2$, but only a small area is amenable to CBM production. This is the area where deep mining is done. Coal seams are generally thick (30–40 ft) and highly inclined. They occur near the boundary between Alberta and British Columbia. The four best prospects are in Horseshoe Canyon, Pembina, Mamille, and Alberta/BC foothills, with a total reserve of about 4–50 TCF. Vertical drilling with hydrofracking can be used for gas production to a depth of 3000 ft. For deeper formations, horizontal drilling from the surface with hydrofracking will be more productive.

The CBM industry in Canada has not reached its full potential, even though 3500 CBM wells have been drilled in Alberta and they are producing 100 BCF (2.5 Bm^3) annually. Projected forecast for CBM production is at 512 BCF/year (14.5 Bm^3) by 2015 [11]. Reservoir characteristics are largely unknown but a gas content of 300 ft^3/t at a depth of 1000 ft was measured near Hinton, BC, in the Jewel seam (internal, unpublished reports by the author). The permeability is about 10 md at this depth. The coal rank is bituminous to low-volatile coking coal.

1.2.3 Western Europe (The United Kingdom, France, and Germany)

These countries have a long history of coal mining. All shallow coal seams are almost mined out. The potential for gas production lies in deeper (3000 ft and deeper) coal seams. However, gas emitted by abandoned mines is being actively collected and utilized.

1.2.3.1 The United Kingdom

There are five major coal producing areas, i.e., Central Valley, northern area, eastern area, western area, and South Wales, with a potential for CBM production. The estimated recoverable gas reserve exceeds 100 TCF. Only a limited effort has been made to drill vertical wells and hydrofrack them but the results are rather disappointing. This technique is likely to succeed only in South Wales, where the geological characteristics of coal seams are favorable. For the rest of the deeper coal deposits, the only technique that has a potential to produce gas at commercial production rates is horizontal BH drilled from the surface with sequential

Table 1.4 Reservoir properties of West European coal deposits

Reservoir properties	Country		
	United Kingdom	France	Germany
Depth (ft)	3000–10,000	3000–10,000	3000–12,000
Gas content (ft^3/t)	100–350	450–500	300–450
Total coal thickness (ft)	100	150	130
Permeability (md)	0.5–1.5	1 (estimate)	1 (estimate)
Reservoir pressure gradient (psi/ft)	0.3–0.4	0.3–0.4	0.3–0.4
Diffusivity (cm^2/s)	10^{-8}	NA	NA
Vertical well production (MCFD)	150	20–100	20–150
Specific gas production (horizontal BH) MCFD/100 ft	3	NA	NA

hydrofracking. The coal seams are generally thin and the permeability is very low. Reservoir properties are shown in Table 1.4 [12].

1.2.3.2 France

The best gas production potential is in the eastern France, the Lorraine–Sarre basin. Coal deposits are deeper than 3000 ft with a total CBM reserve of 15 TCF. Most of the coal is of high volatile bituminous rank and is gassy. The basin covers nearly 3000 $mile^2$. Reservoir properties of coal are shown in Table 1.4 [13].

Based on personal experience, vertical drilling and hydrofracking is not likely to succeed for seams deeper than 3000 ft. Horizontal BH drilled from the surface combined with hydrofracking is likely to succeed.

1.2.3.3 Germany

The estimated recoverable reserve of deep coal is about 183 million tons but about 7 billion tons (mostly lignites) are indicated with a CBM reserve of over 100 TCF. Mining to a depth of 3000–4000 ft was done in the Ruhr and Saar coal basins. The Saar basin has an area of only 440 $mile^2$. Coal seams are of medium thickness with very low permeability. Vertical drilling with hydrofracking was tried but with limited success [14]. Reservoir properties are shown in Table 1.4 [14]. Again, the best method to produce commercial quantities of CBM from these coal seams is to drill horizontal BH from the surface and hydrofrack the horizontal legs. Germany has a large reserve of lignite but it is not a reserve for CBM.

1.2.4 Eastern Europe (Poland, Czech Republic, Ukraine)

1.2.4.1 Poland

The total reserve of hard (and deep) coal is estimated at 100 billion tons with a CBM reserve of 20–60 TCF. The major potential for CBM production is in the upper and lower Silesian basins, which border and extend into Czechoslovakia. Mining has been done to a depth of 3500 ft. Hence all potential reserves of CBM are deeper than that. The gas content of coal seams is high at 635–950 ft^3/t (18–27 m^3/t). Vertical drilling with hydrofracking were planned but results, if any, are not known. The gas production technique that is likely to succeed is horizontal BH drilled from the surface with hydrofracking at 1000 ft intervals.

1.2.4.2 Czech Republic

CBM production potential exists mainly in the upper Silesian basin, also known as the Ostrava-Karvina basin. The basin has an area of 600 $mile^2$ and has many coal seams with a total thickness of about 500 ft. Coal seam gas content is similar to Polish coal fields and is in excess of 700 ft^3/t (20 m^3/t). This provides an excellent opportunity for commercial CBM production using both vertical wells completed in multiple horizons to a depth of 3300 ft (1000 m) and horizontal BH drilled from the surface with hydrofracking at 1000 ft. intervals in deeper coal seams. Preliminary efforts at vertical drilling and hydrofracking by a British firm did not succeed [15] but the process needs to be investigated for improvement.

1.2.4.3 Ukraine

In Ukraine, there are 330 coal seams to a depth of 6000 ft (1800 m) but only 10 are amenable to CBM development [16]. The remaining seams are too thin for commercial exploration. The Donetsk basin (also called Donbass) is the main area of interest. The recoverable coal reserve is estimated at 213 billion tons with a CBM reserve of 63 TCF (1.8 Tm^3). Very little is known about whether any effort to produce CBM commercially has been made. The coal seams are of low vol to high vol bituminous rank and likely to contain 300–600 ft^3/t of CBM. Seam properties are similar to those in the central Appalachian basin of the United States.

1.2.5 Russia

As shown in Table 1.3, Russia has the largest coal reserves and hence the largest CBM reserves in the world. The lower estimate ranges from 2600 to 2800 TCF (75–80 Tm^3). With the abundance of natural gas and oil deposits, Russia has no incentive to produce CBM from its coal deposits. Only about 30% of Russia's coal reserve is of high rank, which is the reservoir for CBM. Areas of interest are in the Donbass (next to Ukraine), Pechora, Karganda, and Kuznetsk basins. The vast majority of Russian coal is of low rank, which does not contain much CBM. Hard coals occur to a depth of 8000 ft (2500 m). Only rudimentary efforts have been made to drill vertical wells with hydrofracking. Four experimental wells drilled to a depth of 2000–3000 ft produced only 35–100 MCFD. This is low compared to US CBM wells. A better technique to produce CBM would be to use horizontal BH drilled from the surface with hydrofracking of the horizontal legs at 1000 ft intervals.

1.2.6 China

China has a vast reserve of CBM, estimated at 1100 TCF (317 Bm^3) to a depth of 6500 ft (2600 m). There are four areas of coal deposit that contain most of the recoverable CBM: (a) Northern (56.3%), (b) North Western (28.1%), (c) Southern (14.3%), and (d) North Eastern (1.3%) (US EPA, 2009). The Chinese CBM industry is off to a good start with some help from the US EPA. More than a thousand vertical CBM wells have been drilled and production enhanced by hydrofracking. Current CBM production is estimated at 130 BCF (4 Mm^3) per year and is increasing. CBM production from shallow minable coal (to a depth of 3000 ft) has become necessary for mine safety. Mine explosions and resulting fatalities are still quite high in China. While vertical drilling with hydrofracking should produce high rates of gas production in all coal fields of China, they must consider using horizontal BH drilled from the surface with hydrofracking for coal seams deeper than 3300 ft (1000 m).

1.2.7 India

Although India has 17 coal fields with a total coal reserve of 200 billion tons, only three basins are viable reserves for CBM, namely Ranigunj (West Bengal), Jharia (Jharkhand), and Singrauli (Madhya Pradesh). The deep coal is of high rank with a gas content of 100–800 ft^3/t. Mines to a depth of 4000 ft (1200 m) producing more than 10 m^3/t (353 ft^3/t) of methane for each ton of raw coal are considered Degree III gassy. The reserves of these mines are good candidates for CBM production. Recent estimates of CBM reserves by Indian agencies put the CBM reserve at 70–100 TCF (2–3.4 Tm^3) [17]. Four blocks of coal covering an area of about 6000 $mile^2$ have been leased for CBM production. Over 100 wells have been completed but production data is not yet available. These are vertical wells completed in a single coal seam with hydrofracking. Coal seams are generally thin at greater depth. Some very thick seams (in Bihar and Odhisa provinces) at shallow depths may turn out to be good producers (like the Powder River basin of the United States) but attempts to produce CBM have not yet been made.

1.2.8 South Africa

The central part of the country containing the Witbank and Highfield basins is the best prospect for CBM production. Most of South Africa's coal is produced from these two basins. The coal is of high rank but most coal seams are shallow. The average gas content is estimated at 300 ft^3/t at a depth of 1000 ft. Preliminary efforts to drain methane by in-mine horizontal BH and vertical wells with hydrofracking are afoot but results are not available yet [18]. The estimated CBM reserve is low, at 5–10 TCF. Geological conditions of these basins (too shallow) preclude the use of vertical drilling and hydrofracking. Commercial gas production can only be obtained if horizontal BH are drilled from the surface and have a lateral extension of 3000–5000 ft. CBM production techniques used in the Northern Appalachian basin of the United States have a potential application in South Africa. Some methane gas has been captured from gold mines and used for many years but the subject is beyond the scope of this book.

1.2.9 Australia

The best prospect of CBM production lies in the Bowen basin (Queensland) and the Sydney basin (New South Wales). The latest estimate of CBM in these basins is about 7 TCF. The Sydney basin coal seams are deeper and gassier with a gas content of 350–700 ft^3/t (10–20 m^3/t) and, as such, more amenable to gas production by vertical wells with hydrofracking. In-mine horizontal drilling has shown fairly good specific gas production of 8–10 MCFD/100 ft of BH. Well-designed wells with multiple completion in several coal seams can produce 200–300 MCFD per well. Bowen basin coal seams have been extensively drilled, with over 500 wells. Annual gas production from this basin is estimated at 100 BCF which is about 90% of total Australian CBM production. Vertical drilling and hydrofracking is unlikely to be very productive in the Bowen basin because of its shallow depth. Horizontal drilling from the surface is the best technique to produce gas from these shallow coal seams. Hydrofracking may not be needed but actual gas production from such BH will dictate it. However, even the current low CBM production is providing up to 48% of Queensland's gas supply [7].

1.2.10 Other Coal-Producing Countries

Even though there is a general lack of reliable data, there are potentially good CBM reserves in many countries other than those listed above. The ideal way to locate and prove a CBM reserve is to drill the area on a grid pattern, collect cores, and measure the gas content. Where such data are not available, one can use some general guidelines to locate a potential CBM reserve.

1. All deep, thick seams of high rank coal are potential CBM reserves.
2. Very thick coal seams (100–300 ft) can be good reserves even if they are shallow and of low rank. The Powder River basin is a good example.
3. The reserves of coal mines that have high specific methane emissions are also a potential reserve. The specific methane emission of a mine is the volume of methane produced per ton of raw coal mined. It is linearly related to the depth of the coal seam as shown in Fig. 1.3 [19]. Coal reserves where mines have a specific methane emissions of more than 700 ft^3/t (20 m^3/t) are potentially good reserves.

In conclusion, it must be noted that even though there is vast potential for CBM production worldwide, production growth is controlled by a number of nontechnical factors. The main three are:

1. The current over-supply of natural gas in the world.
2. Lack of equipment and technology outside the United States for horizontal completions from the surface.
3. Environmental laws, particularly in Europe, driving the cost very high. A typical water injection well drilled in Eastern Europe will cost three times more than a similar well in the Central Appalachian basin of the United States.

The drilling and completion techniques as well as the reservoir and production engineering for CBM are significantly different from those for

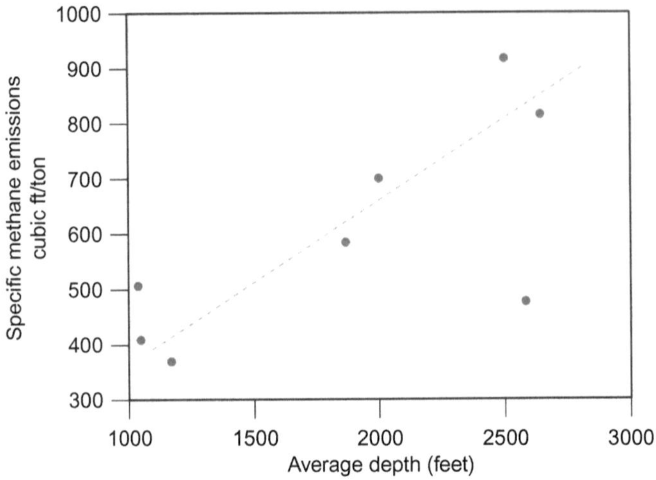

Figure 1.3 Underground mines' specific methane emissions versus depth.

Table 1.5 Comparison of CBM and natural gas reservoirs

CBM reservoir	Natural gas reservoir
Gas adsorbed on the surface of microscopic coal particles	Gas stored in the pores of the host rock
Gas flows first by diffusion from coal (Fick's law) and then through fractures in coal (Darcy's law)	Strictly Darcy flow
Gas content/ton of coal high	Gas content/ton of host rock low
Permeability is depth-dependent	Permeability is independent of depth
Sorption characteristics generally limit the gas recovery	Gas recovery is only dependent on reservoir pressure depletion
The reservoir is a carboniferous rock (with a carbon content ≥ 50%)	The reservoir is a noncarboniferous rock
CBM is syngenetic in origin with coal	Gas may not originate in the reservoir

conventional natural gas wells drilled in sandstone and limestone. Table 1.5 shows some critical differences that justify the need for this book.

Gas production from coal, therefore, is a far more complicated process than that for natural gas from sandstone/limestone reservoirs. The purpose of this book is to present reservoir and production engineering for CBM in a simple, understandable language. The knowledge contained in the book will help optimize gas production from all the above-listed prominent CBM reserves.

REFERENCES

[1] Hargraves AJ. Planning and operation of gaseous mines. CIM Bull March, 1973.
[2] World Energy Reserves and Consumption. <http://en.wikipedia.org/worldenergyreserve&consumption/>; 2013.
[3] Landis ER, Weaver JW. Global coal occurrences: hydrocarbons from coal. AAPG Stud Geol 1993;38:1−12.
[4] World Coal Statistics. World Coal Association, <http://worldcoal.org/>; 2013.
[5] The International Coal Seam Gas Report. Cairn point publishing company, Queensland, Australia; 1997.
[6] Kuuskraa VA, Boyer CM, Kelafant JA. Coal bed gas: hunt for quality gas abroad. Oil Gas J 1992;90:49−54.
[7] Global Overview of CMM Opportunities. US Environmental Protection Agency (US EPA) Coal bed methane outreach program, 2009, 260.
[8] Thakur PC. Coal bed methane production. Chapter 11.6 in SME Mine Engineer's Handbook, 2011.
[9] Thakur PC, et al. Coal bed methane: from prospect to pipeline. Amsterdam: Elsevier; 2014. p. 420.
[10] Rogers, et al. Coal bed methane; principles and practices. Starkville: Oktibbeha Publishing; 2011.
[11] NAEWG, North American Natural Gas Vision Report. North American Energy working group experts group on natural gas trade and interconnections, <www.pi.energy.gog/pdf/library/NAEWGGASVISION2005.pdf>; 2005.
[12] Creedy DP. Prospects for coal bed methane in Britain, Coal Bed Methane Extraction Conference, London, 1994. 1−18.
[13] Gairaud H. Coal bed methane resources and current exploration/production work Coal Bed Methane Extraction Conference, London; 1994. 1−26.
[14] Schloenbach M. Coal bed methane resources of Germany's Saar basin and current activities. Coal Bed Methane Extraction Conference, London; 1994
[15] BERR/DTI, Assessment of cleaner coal technology market opportunities in Central and Eastern Europe, Department of Business and Regulatory Reform, Project report R270, DTI/Pub URN 04/1852; 2004.
[16] Lakinov V. Prospects for CBM industry development in Ukraine, M2M workshop—Ukraine, Beiging, China; 2005.
[17] DGH, CBM exploration, Directorate general of hydrocarbons. Ministry of Petroleum and Natural Gas; New Delhi, India; 2008.
[18] UN FECC, South Africa's initial national communication to UN framework convention on climate change. p. 77. <http://unfecg.int/resource/docs/nate/zafnc01.pdf>; 2000.
[19] Thakur PC, et al. Global methane emissions from the world coal industry. Proceedings of the International Symposium on Non-CO2 Green House Gases, Why and How to Control, Maastricht; The Netherlands; 1993. p. 73−91.

CHAPTER 2

Gas Content of Coal and Reserve Estimates

Contents

Abstract

Gas content of a coal seam is perhaps the most important reservoir property. Techniques to measure the gas content are discussed. A gas "isotherm" for a coal seam is another important characteristic that shows the relationship between gas content and reservoir pressure. Mathematical derivation of this relationship (for indirect estimation of gas content) is done and gas isotherms with key parameters are derived for some prominent US coal seams. Influences of various factors, such as depth/reservoir pressure, rank of coal, temperature, moisture, and ash content of coal, are discussed. Reserve estimates of minable coal seams but particularly, the gas production from overlying and underlying coal seams is also presented. A procedure to estimate the gas reserve of deeper/nonminable coal seams is also presented. It involves coring, gas content measurement in the laboratory, and proximate analysis of coal. Finally, important properties of most components of coal bed methane, such as methane, ethane, propane, hydrogen, carbon dioxide, and nitrogen, are listed.

Advanced Reservoir and Production Engineering for Coal Bed Methane.
DOI: http://dx.doi.org/10.1016/B978-0-12-803095-0.00002-8

The most important reservoir properties that not only influence the gas production rates but also determine the production techniques are:

- Gas content of coal and its gas isotherm;
- Matrix permeability;
- Depth and reservoir pressure;
- Diffusivity of coal;
- Water content and quality of water;
- Ground stress; and
- Elastic properties of coal and surrounding strata.

The volume of gas contained in coal at standard temperature and pressure (STP)[1] is termed the *gas content* of the coal and is expressed in cubic feet per ton. It is generally accepted that gas is stored in a monolayer on the microscopic particles of coal that are smaller than the micropores in the coal matrix. At greater depth, the gas may be in a "condensed, liquid like state" [1]. The volume of gas retained in coal is dependent on the rank, temperature, and pressure or depth of the coal seam. The microscopic surface of coal is large; a ton of coal has a surface area of approximately 2218 million ft^2 (200 Mm^2). Thus one cubic foot of coal can store two to three times the amount of gas contained in a typical sandstone reservoir for natural gas of the same volume but at higher pressure. For commercial gas production, it is best to core drill the entire reserve on a grid pattern (typically 1 core hole per 500 acres) and do a direct measurement of the gas content of all coal seams that make up the gas reservoir.

Gas content measurement methods are classified as (a) conventional, and (b) pressurized desorption techniques. In the conventional technique, coal cores or drill cuttings are retrieved from the core holes and immediately put in a sealed container to measure the desorbed gas. This method suffers from uncertainty in the estimate of gas lost during sample retrieval and handling. To eliminate this problem, the pressurized core desorption technique has been developed. In this technique, gas loss is minimized by sealing the coal samples while they are in the core hole. Both methods provide positive proof of gas presence. Data on gas desorption rate can be used for the calculation of diffusivity (to be discussed later) and to determine if the coal is liable to instantaneous outburst in mines. Desorbed gases are chemically analyzed to determine the composition and calorific value of coal bed methane (CBM).

[1] STP means 32°F and 14.7 psi.

2.1 THE DIRECT METHOD OF GAS CONTENT MEASUREMENT

This technique was originally developed by Bertard and Kissell [2,3]. It was further improved by Diamond and Schatzel [4] and became the "ASTM standard practice for determination of gas content of coal" (ASTM D 7,569-10, 2010) [5]. In this technique, the desorbed gas from the coal sample is measured first. Next the cumulative gas production is plotted against the square root of time to determine the lost gas. Finally, a small, weighted portion of coal sample is crushed in a hermetically sealed mill to get the residual gas. The total gas content is the sum of the three components: (a) desorbed gas, (b) estimated lost gas, and (c) residual gas.

2.1.1 Desorbed Gas

After coal cores or drill cuttings are put in a hermetically sealed container, called a desorption canister, the desorbed gas is measured periodically. In the first few days, readings may be taken every hour, but later a measurement once a day is sufficient. The general layout of the experimental set up is as shown in Fig. 2.1 [4].

The desorption canister is about 18 inches tall, with a 4 inch internal diameter. It is equipped with a pressure gage and a valve to let the desorbed gas out. The desorbed gas is measured by water displacement in a graduated glass cylinder 4 inches in diameter and 12 inches high. The glass cylinder is connected to a leveling water reservoir, and the gas volume measurement is taken when the water levels in the cylinder and leveling reservoir are the same. The precision of the measurement is about ± 4% [6].

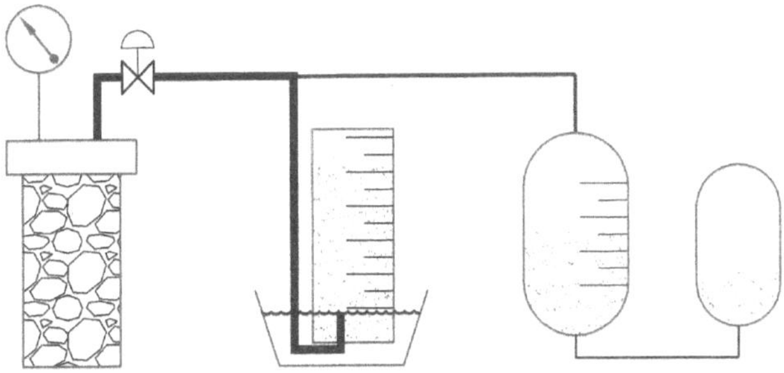

Figure 2.1 Gas content measuring apparatus.

The desorption process typically extends to 4–6 weeks. It is stopped when gas desorption is less than 10 cm^3/day. The cumulative gas production is plotted on a graph paper against $(\text{time})^{0.5}$ to determine the lost gas component of the total gas content (discussed in the next paragraph). The desorbed gas is periodically analyzed using a gas chromatograph to determine its composition and calorific value.

2.1.2 Lost Gas

This portion of the total gas content is the gas that escapes from the sample during its collection and retrieval, prior to being sealed in an air-tight canister. It is estimated indirectly. Most gas desorption processes from coal or shale follow a power law [7,8].

$$Q = At^n \tag{2.1}$$

where

Q is the cumulative volume of gas desorbed in ft^3
A is a characteristic of the coal (equals initial production in gas wells)
t is time in days or minutes
n is a characteristic of the coal or shale

Eq. (2.1) can be expressed in its logarithmic form as

$$\ln Q = \ln A + n \ln t \tag{2.2}$$

The value of "n" for most coal is 0.8–1.00. Hence a plot of $\ln Q$ against $\ln t$ yields a straight line. The intercept on the "y" axis is equal to $\ln A$.

In a simplified version of Eq. (2.1),

$$Q = B + nt^{0.5} \tag{2.3}$$

Hence a plot of cumulative desorbed gas, Q, against $(t)^{0.5}$ yields a straight line. Here B is the intercept on the y axis and is a measure of the lost gas as shown in Fig. 2.2. The value of "n" here is about 1.00.

2.1.3 Residual Gas

Even when the coal sample in the desorption container has stopped producing gas, a significant volume of gas is still left in the sample. It can only be retrieved and measured by crushing the sample to very fine sizes.

A hermetically sealed modified ball mill (Bleuler Mill) [9] is used for this purpose. A measured quantity of the coal core or drill cutting is put in the mill and crushed. The released gas is measured by the same setup that was used for desorbed gas measurements.

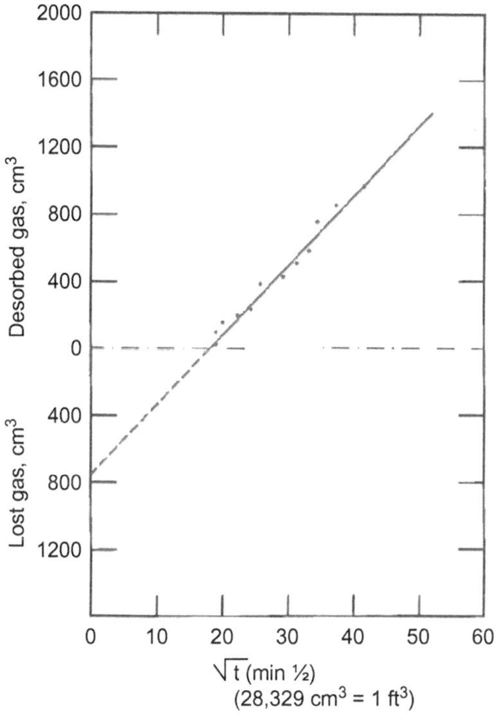

Figure 2.2 Lost gas estimation graph.

The total gas content of the coal sample is obtained by adding the three components, i.e., desorbed gas, lost gas and residual gas. The coal sample is next weighed and sent to a laboratory for a proximate analysis which yields the moisture, ash, volatile matter, and fixed carbon contents of coal. The weight of coal is calculated on a dry, ash-free basis. The total gas content of the coal sample is divided by the weight of the coal sample (dry, ash-free) to get the final gas content of coal, in ft^3/t ($1\ cm^3/gm = 32\ ft^3/t$).

Table 2.1 shows the gas content and composition data for some typical US coal seams.

2.2 GAS ISOTHERMS AND INDIRECT METHODS OF GAS CONTENT DETERMINATION

At constant temperature, each coal seam shows a measurable relationship between the total gas adsorbed (or desorbed) and the confining pressure. Fig. 2.3 shows typical gas isotherms for five US coal seams.

Table 2.1 CBM content and composition of US coal seams

Coal seam	Rank	Gas content (ft^3/t)	Composition[a] (%)					Calorific value BTU/ft^3
			CH_4	C_2H_6	C_3H_8	H_2	CO_2	
Pocahontas #3 (VA)	L.V.	450–650	97–98	1–2	Trace	.02	0.2–0.5	949–1058
Hartshorne (OK)	L.V.	200–500	99.20	0.01	–	–	–	900–1058
Kittanning (PA)	L.V.	200–300	95–98	0.02	Trace	–	0.1–0.2	1020
Mary Lee (AL)	L.V.	200–500	96	0.01	–	–	–	1024
Pittsburgh #8 (WV)	HVA	100–250	89–95	0.25–0.5	Trace	–	2–11	949–1000
Mesa Verde Formation (NM)	Sub bit.	100–300	88	–	–	–	12	938

[a]N_2 and argon contents are not listed but are needed to make the total 100%.

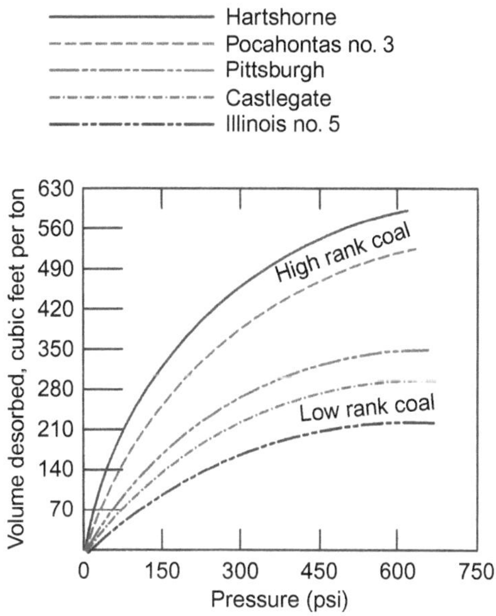

Figure 2.3 Gas isotherms for US coal seams.

It is to be noted that high rank (low vol. bituminous) coals contain more gas than the lower rank (high vol. bituminous) coals (HVA, HVB, HVC) at the same confining pressure. It is also clear that the sorption capacity of all coal increases with pressure, but the increase occurs at an ever- decreasing rate as the sorption capacity reaches an asymptotic limit—the saturation limit.

There are two main mathematical representations for these isotherms:

a. The Langmuir isotherm [10] is expressed as

$$V = V_m \frac{bP}{1 + bP} \tag{2.4}$$

where

V is the volume of gas contained at pressure P, ft^3/t

V_m is the maximum sorption capacity of coal, ft^3/t

P is the pressure, psi

b is the Langmuir constant, psi^{-1}

b. The Freundlich isotherm [11] is expressed as

$$V = mP^k \tag{2.5}$$

where

m and k are characteristic constants of coal

P is pressure, psi

V is volume of gas/ton of coal

In the United States, Eq. (2.4) is most commonly used.

For indirect determination of the gas content of coal at a given pressure, Eq. (2.1) can be rewritten as

$$\frac{P}{V} = \frac{P}{V_m} + \frac{1}{b\,V_m}. \tag{2.6}$$

The term "b" is experimentally found to be equal to $1/P_L$, where P_L is the characteristic pressure that corresponds to $V_m/2$ on the gas isotherm.

Thus Eq. (2.6) can be written as

$$\frac{P}{V} = \frac{P_L}{V_m} + \frac{P}{V_m}. \tag{2.7}$$

If the isotherms shown in Fig. 2.3 are replotted with P/V on the y axis and P on the x axis, a straight line is obtained. The plot for the Pocahontas #3 seam is shown in Fig. 2.4. The slope of the line is $1/V_m$, from which V_m can be determined. The intercept on the y axis is P_L/V_m, from which P_L can be determined.

Calculated values of P_L and V_L for all gas isotherms shown in Fig. 2.3 are shown in Table 2.2.

Some interesting conclusions can be drawn from data in Table 2.2 and the existing gas content and reservoir pressures of these coal seams. The Hartshorne and Pocahontas coal seams are deep (1500–2500 ft depth). Their gas contents are 550–650 ft^3/t and reservoir pressures are 500–650 psi. This indicates that these coal seams are still near their

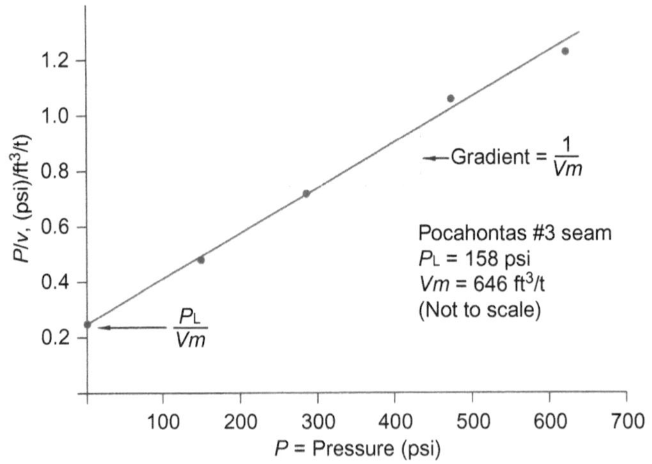

Figure 2.4 A plot of *P/V* against *P*.

Table 2.2 Calculated V_L and P_L values for US coal seams

Coal seam	V_L (ft^3/t)	P_L (psi)
Hartshorne	788	205
Pocahontas #3	646	158
Pittsburgh	443	170
Castlegate	409	229
Illinois #6	353	273

saturation points and have not lost much gas. They are potentially good reserves and will yield high rates of gas production as they have in the past. The Pittsburgh and Illinois coal seams, on the other hand, are relatively shallow (at 1000 ft depth). The measured gas content is typically 100–200 ft^3/t and the reservoir pressure is less than 200 psi. This shows that a considerable amount (50–60%) of their original gas content has been lost and, therefore, these reservoirs would be low producers. These observations are, in fact, confirmed by actual gas production data.

2.3 CALCULATION OF GAS CONTAINED IN THE PORES OF COAL

The indirect method calculates only the adsorbed gases in coal. The gas contained in the fractures and pores in a ton of coal must be added to the indirectly estimated gas content. The porosity of coal seams ranges from 1% to 5%. The general formula for this is

$$V\phi = \phi V_c \times \frac{P}{P_o} \times \frac{273}{T} \tag{2.8}$$

where

$V\phi$ is the volume of gas in pores, ft^3/t at STP
ϕ is the porosity, %
P is the reservoir pressure, psi
P_o is atmospheric pressure
T is reservoir temperature in Kelvin
V_c is the volume of coal per ton

Assuming $\phi = 3\%$
$P = 600$ psi
$P_o = 14.7$ psi
$T = 333\,°K$
$V_c = 25$ ft^3/t

$$V\phi = \frac{0.03 \times 25 \times 600 \times 273}{14.7 \times 333} = 25 \text{ ft}^3$$

For an indirectly estimated gas content of 500 ft^3/t, the pore gas will contribute 5% additional gas.

2.4 INFLUENCE OF VARIOUS PARAMETERS ON THE GAS CONTENTS OF COAL

Only the main parameters that influence the gas content of coal are discussed.

2.4.1 The Reservoir Pressure

This is by far the most important factor. In general, the deeper the coal seam, the higher the reservoir pressure and the gas content. Fig. 2.3 shows this clearly.

2.4.2 Rank of Coal

Fig. 2.3 also shows the isotherms of various ranks of coal. In general, the higher the rank of coal, the higher is the gas content. Hartshorne and Pocahontas #3 are low vol. metallurgical coal of high rank, but Pittsburgh, Castelgate, and Illinois #6 are progressively lower rank coals.

2.4.3 Temperature

Temperature has the opposite effect to pressure. As the temperature of coal formation increases, the adsorbed gas volume/ton decreases. The gas content reduces by 0.8% for each one degree Celsius rise in temperature for bituminous coal [12].

2.4.4 Moisture

Moisture tends to decrease the gas content of coal. A rough estimate is provided by the following equation [13].

$$\frac{V_{\text{moist}}}{V_{\text{dry}}} = \frac{1}{1 + 0.31\ W} \tag{2.9}$$

where

W is the moisture content, %

Thus a 5% increase in moisture content in coal can reduce the gas content of dry coal by nearly 60%.

2.4.5 Ash

The gas content of pure coal appears to slightly increase as the ash (or mineral matter) content of coal reduces. The relationship is expressed as

$$\frac{\text{Gas pure coal}}{\text{Gas dirty coal}} = \frac{1}{1 - 0.01\,A} \tag{2.10}$$

where

A is the ash content in percent (usually less than 50%) [12].

Gas content is usually reported on a dry, ash-free basis.

2.5 GAS RESERVE ESTIMATION

CBM production can be realized from two types of reserve, i.e., (a) mineable coal seams with operating mines and (b) nonmineable areas.

2.5.1 Reserve Estimation for Mineable Areas

Besides the gas recovered from the virgin coal to be mined, the mining activity can create a vast gas emission space by causing the overlying gas-bearing strata to subside and heaving the floor with additional gas-bearing strata.

Fig. 2.5 shows the vertical limit of the "gas emission space" [14]. All the gas released by the gas emission space goes to the mine because it acts like a pressure sink. Typically coal is mined by removing a slice of coal seam measuring 10,000 ft × 1200 ft, equal to 275 acres at a time. The total gas recovered from this mined area is measured and divided by the acres mined to get the "specific gob emissions" for the area. For example, the number is 30 MMCF/acre for the Pocahontas #3 coal seam in central Appalachia.

The next step is to multiply the total mineable acres by the "specific gob emission" to determine the recoverable gas reserve. A typical mine may have an area of 100,000 to 200,000 acres with a life of 30–60 years. Assuming 50% coal recovery, the gas reserve for a large mine like this is thus 1.5–3 TCF. On a mining property like this, vertical wells are drilled 10 years ahead of mining. Assuming the mine has 500 wells drilled in advance of mining, producing on an average 300 MCFD, the total gas production is equal to 150 MMCFD. A typical longwall face mining at 1 acre/day will add another 30 MMCFD. Thus a daily production of 180 MMCFD is attainable. This is considered a viable commercial production rate by the gas industry.

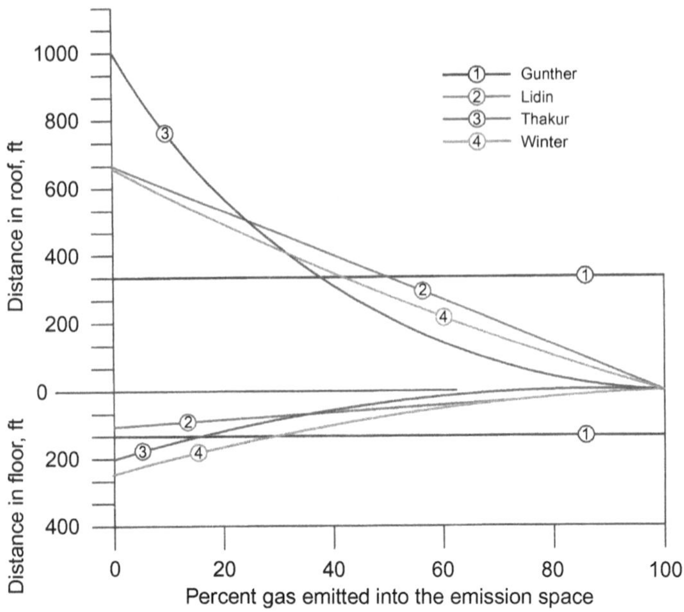

Figure 2.5 Vertical limits of the gas emission space.

2.5.2 Gas Reserve Estimation for Nonmineable Reserve

This is a simple mathematical procedure. In a typical CBM reservoir, there are many (often more than 30) coal seams to a depth of 10,000 ft, but only a few will be amenable to commercial gas production. Two main production techniques (to be discussed later) have limitations that restrict the reserve volumes.

a. Vertical wells with hydrofracking: This technique works at a depth of 1500–3000 ft. In wells shallower than 1500 ft, the fracking process is inefficient (horizontal fractures). In wells deeper than 3000 ft, the permeability is usually too low for good gas production. Typically, coal seams thinner than three feet are not fracked. These criteria will help in selecting the target coal seams for gas production.

b. Horizontal wells drilled from the surface: This technique works at all depths but is usually limited to only one thick coal seam to keep the cost within limits. In shallow coal, the horizontal laterals are about 3000 ft each. In deep coal, the horizontal laterals can reach 5000–10,000 ft.

The gas reserve of each coal seam targeted for production is estimated separately and the estimates are added to get the total gas reserve.

Table 2.3 Property of gases in CBM

Compound	Methane	Ethane	Propane	Hydrogen	Carbon dioxide	Nitrogen	Air
Formula	CH_4	C_2H_6	C_3H_8	H_2	CO_2	H_2	N_2-O_2
Molecular weight	16.04	30.06	44.09	2.02	44.01	28.02	28.97
Inverse density at STP (ft^3/lb)	23.61	12.52	8.47	188.67	11.05	13.53	13.09
Specific gravity	0.55	0.41	1.02	0.069	1.55	0.97	1.00
Viscosity (micro poise) at 0°C	202.6 (at 380°F)	34.8	75	83.5	136.1	165	178.8
Critical pressure (psi)	673	708.3	617.4	188	1073	492	547
Critical temperature (°F)	−116.5	90.09	206.26	−399.8	88	−232.8	−221.3
Specific heat at STP (Cp), BTU/lb./degree F	0.53	0.39	0.34	3.34	0.20	0.25	1.24
Gross calorific value, BTU/ft^3 at STP	1012	1783	2557	324	—	—	—

For a single seam the gas reserve G is given as

$$G = 43,560\ A\ H \times C_g\ \text{ft}^3 \tag{2.11}$$

where

A, acre in acres

(1 acre = 43,560 ft^2)

H = height of coal seam in feet

C_g = gas content in ft^3/ft^3 of coal

(1 ton of coal is typically 25 ft^3 in volume)

Thus, for an area of 100,000 acres in a coal seam of 6 ft height with a C_g of 25 ft^3/ft^3 of coal, the gas in place (GIP) = 43,560 × 100,000 × 6 × 25 = 653 BCF.

It is to be noticed that all the GIP will not be recovered. Coal properties, mainly diffusivity, determine the recovery rate. In coals with low diffusivity (less than 10^{-8} seconds^{-1}), the recovery may be only 60%, but for coal seams with high diffusivity (10^{-6} seconds^{-1} and above), the recovery may be as high as 80%. These recovery factors were actually obtained in US coal seams of the Appalachian region by measuring the gas content of the virgin coal and that of the totally degassed coal after years of gas production.

2.6 PROPERTIES OF THE CBM PRODUCED

Before marketing, the CBM must be processed to meet commercial pipeline requirements. The specifications vary but typically a BTU of 960 with no more than 4% of noncombustibles ($N_2 + CO_2$) is required. Oxygen is limited to 0.2–1.00 ppmv and moisture content should not exceed 7 lb/MMCF of gas.

Table 2.3 shows the physical properties of the most important gases in CBM (adapted from the *Physics and Chemistry Handbook* [15] and the *Handbook of Natural Gas Engineering* [16]).

REFERENCES

[1] Yee D, et al. In: Law RE, Rice DD, editors. Gas sorption on coal and measurement of gas content, in hydrocarbons from coal. Tulsa: AAPG; 1993. p. 159–84.

[2] Bertard C, et al. Determination of desorbable gas concentration of coal. Int J Rock Mech Min Sci 1970;7:43–65.

[3] Kissell FN. The direct method of determining methane content of coal beds for ventilation design, US Bureau of Mines, RI 7767, 1973. p. 17.

[4] Diamond WP, Schatzel SJ. Measuring the gas content of coal: a review. Coal Geol 1998;35:311–31.
[5] ASTM Designation: D7569-10. Standard practice for determination of gas content of coal—direct desorption method; 2010.
[6] TRW. Desorbed gas measurement system—design and application. US Department of Energy Contract No. DE-AC21-78MC08089, METC, Morgantown, WV; 1981.
[7] Thakur PC. Methane control on longwall gobs, Longwall-shortwall mining, state of the art. AIME; 1981. p. 81–6.
[8] Thakur PC. Methane flow in Pittsburgh coal seam. Harrogate: The 3rd International Mine Ventilation Congress; 1984. p. 177–82.
[9] Thakur PC. Mass distribution, percent yield, non-settling sizes and aerodynamic shape factors of respirable coal dust particles, MS Thesis, Penn State University; 1971. p. 133.
[10] Langmuir I. The adsorption of gases on plane surfaces of glass, mica and platinum. J Am Chem Soc 1918; vol. 40:1361–403.
[11] Yang RT, et al. Gas separation by adsorption process. Boston: Butterworth Publishers; 1987.
[12] Boxho J, et al. Fire damp drainage, VGE. Essen: Verlag Gmbh; 2009. p. 419.
[13] Ettinger, et al. Systematic handbook for the determination of methane content of coal seams from the seam pressure of the gas and methane capacity of coal. Institute of Mining, Academy of Science, USSR, USBM translation 1505; 1958.
[14] Thakur PC, Zachwieja J. Methane control and ventilation for 1000 ft wide longwall faces. USA: Longwall; 2001. p. 167–80.
[15] Hodgman CD, editor. Handbook of chemistry and physics. Cleveland, OH: The Chemical Rubber Publishing Company; 1941. p. 2324–5.
[16] Katz DL. Handbook of natural gas engineering. Newyork: McGraw-Hill Book Company; 1959. p. 131–2.

CHAPTER 3

Porosity and Permeability of Coal

Contents

Abstract

Next to gas content, permeability is the most important coal bed methane reservoir property. Porosity is a less important parameter for coal. Both porosity and permeability are defined. An experimental technique for measuring porosity is discussed. Three methods to measure permeability are discussed: (a) theoretical, (b) experimental, and (c) field measurements. The theoretical method assumes coal to have a well-defined matrix structure, which is not always true for an anisotropic rock like coal. Standard laboratory techniques, such as the Gas Research Institute and pulse decay techniques, are discussed. They usually underestimate the real permeability. Various field measurement techniques are listed, but only three are discussed in detail. The most common quick technique involves a "mini-frack" of the coal formation. It yields much more information about the coal seam in addition to permeability.
Derivation of permeability from "closure pressure" is discussed. Two well-known techniques for field measurement of permeability are discussed: the draw down and the build-up tests. Data for typical test runs are plotted and used to calculate the effective permeability of the coal seam. These measurements of permeability are the most accurate values. Finally, the influence of depth and coal seam temperature, and shrinkage of the coal matrix, including the Klinkenberg effect, on permeability is discussed.

Advanced Reservoir and Production Engineering for Coal Bed Methane.
DOI: http://dx.doi.org/10.1016/B978-0-12-803095-0.00003-X

For successful analysis of fluid flow in a porous medium, such as coal, it is important to determine two basic properties of the reservoir; porosity and permeability.

3.1 DEFINITION OF POROSITY

Porosity is the fraction of the total volume of a rock that can hold gas or liquid. In other words, it is the percentage of the bulk volume of the rock that is not occupied by solid matter. Coal seams are a fractured matrix. Fig. 3.1 shows the two main macrofractures in the coal matrix, called the face cleat and the butt cleat. One cleat direction is orthogonal to the other. The face cleat is the major fracture that stores and conducts gases. The butt cleat is the minor fracture. The space contained in these fractures comprises the majority of the porosity of coal.

Mathematically porosity can be expressed as

$$\phi = \frac{V_p}{V_b} = \frac{V_b - V_m}{V_b} \tag{3.1}$$

where:

V_p is connected pore volume
V_b is the bulk volume
V_m, is the volume of the solid matrix material

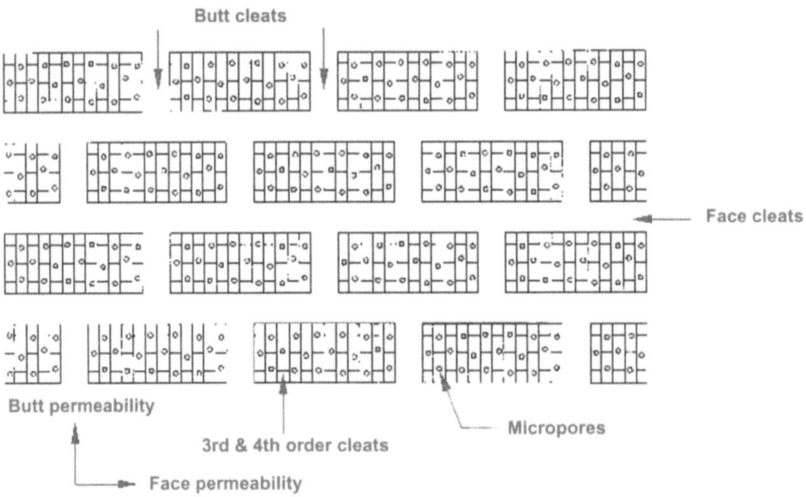

Figure 3.1 A simplified coal matrix.

Porosity can be determined by measuring any two of the three parameters in the above equation.

The major fractures in coal, also called macropores, are essential for the transport of methane and water but are not significant for methane storage, as discussed in Chapter 2, Gas Content of Coal and Reserve Estimates. The porosity of the cleat system in US coals ranges from 1% to 5% [1].

3.2 MEASUREMENT OF POROSITY

A specimen of the coal/rock about one inch in diameter and one to two inches long is prepared. The volume of the solid matrix is determined by

1. Weighing the core plug,
2. Placing the plug in a vessel and evacuating the air,
3. Admitting a liquid of known density (e.g., tetrachloroethane) to submerge the plug and returning to atmospheric pressure, and
4. Weighing the saturated plug, taking care to remove all extraneous fluid.

The bulk volume of the specimen can be determined by fluid displacement in a pycnometer.

$$\text{Porosity, \%} = \frac{\text{Volume of pores} \times 100}{\text{Volume of core}} = \frac{\text{Weight of liquid in pores} \times 100}{(\text{density of liquid})(\text{volume of core})} \tag{3.2}$$

For example:

volume of core = 9 cc

weight of dry core = 21 gm

weight of core saturated with tetrachloroethane = 22.3 gm

weight of liquid in pores = 1.3 gm; density of tetrachloroethane = 1.6 gm/cc

$$\text{hence: } \phi = \frac{1.3 \times 100}{1.6 \times 9} = 9\%$$

As discussed in Chapter 2, Gas Content of Coal and Reserve Estimates, the fractures in coal are essential for gas transport but they do not store much gas. The blocks of coal created by the intersection of the two cleat systems have micropores composed of capillaries and cavities of molecular dimensions. This is where most of the gas in coal is stored, by adsorption on the vast surfaces of the coal. Gas contained in the macropores and micropores of coal is significant when compared to other low-permeability reservoirs such as shale, but constitutes only 5–10% of the total gas content of coal. How much depends on the rank of coal and the depth of the coal seam, among other variables.

3.3 DEFINITION OF PERMEABILITY

Permeability is a property of a porous rock such as coal and is a measure of the capacity of the medium to transmit fluids. It depends on the driving pressure differential, the area of the specimen, and the viscosity of the fluid.

Mathematically, it can be written as

$$u = \frac{Q}{A} = -\frac{k}{\mu}\frac{dp}{dx} \tag{3.3}$$

where

u is the average fluid velocity $= \frac{Q}{A}$ in cm/second
A is the cross-sectional area in cm^2
k is the permeability of the medium in darcy
μ is the viscosity of gas/liquid in centipoise
$\frac{dp}{dx}$ is the pressure gradient in atm/cm

A negative sign indicates that fluid flows in the direction of the declining pressure gradient. Since most mineable coal seams are shallow (less than 3000 ft in depth), the fluid can be assumed to be noncompressible. Integrating Eq. (3.3) for the length of the specimen, L,

$$\frac{Q}{A}\int_{o}^{L} dx = \frac{-k}{\mu}\int_{P_2}^{P_1} dp \tag{3.4}$$

or

$$Q = \frac{kA}{\mu L}(P_1 - P_2) \tag{3.5}$$

In an experiment to measure k, all the parameters in Eq. (3.5) are known and hence permeability can be easily determined.

A cube of coal 1 cm on a side will have a permeability of 1 darcy, if a fluid of 1 cp viscosity flows between the back and front faces of the cube at a rate of 1 cc/second under a pressure differential of 1 atmosphere at 68°F. Converted to SI units, 1 darcy is equivalent to $9.869233 \times 10^{-13}\ m^2$ or roughly 1 mm^2. Since a darcy is a very large unit, the permeability is mostly expressed in 1/1000 of a darcy or millidarcy (md) and it has a dimension of L^2.

The above equation is valid for liquids. For gases, the volume q is introduced as defined by

$$q = Q \cdot \frac{P_1 + P_2}{2P_b} \tag{3.6}$$

Substituting in Eq. (3.5) and expressing k in md, the equation for gas flow can be written as

$$k = \frac{2000\, q\, L\, \mu P_b}{A(P_1^2 - P_2^2)} \tag{3.7}$$

where

k = permeability in millidarcy
q = gas flow rate in cm^3/second
L = length of the specimen in cm
μ = gas viscosity in centipoise
P = absolute pressure in atm
subscript 1 = upstream core
subscript 2 = downstream core
b = base pressure of gas measurement.
For example, assume
$q = 2$ cm^3/second
$P_1 = 2$ atm
$P_2 = 1$ atm
$L = 2$ cm; $A = 3$ cm^2
$P_b = 1.00$ atm
$\mu = 0.018$ cp at 68°F.

$$k = 1\left(\frac{2000 \times 2 \times 0.018}{3(4-1)}\right) \times \frac{1.0}{1.0} = 8 \text{ md}$$

3.4 MEASUREMENT OF PERMEABILITY

All techniques for measuring the permeability of a nonhomogenous, viscoelastic, sorptive material such as coal can be divided into three categories:

a. Theoretical calculation of porosity and permeability.
b. Laboratory-based techniques. Usually the same apparatus is used to measure both porosity and permeability.
c. Field-based techniques for effective permeability.

3.4.1 Theoretical Calculation of Porosity and Permeability

Robertson and Christensen [2] developed a cubic geometry to simulate the coal matrix. The major fracture was assigned a width of "*b*," whereas the fractureless coal cube has a side equal to "*a*."

Fig. 3.2 shows the schematic of this idealized system. Mathematical calculations can be done to show that

$$\text{Porosity}, \phi = \frac{3\,b}{a} \tag{3.8}$$

Likewise, the permeability was calculated as

$$k = \frac{b^3}{12\,a}(b << a) \text{ darcy} \tag{3.9}$$

"*a*" and "*b*" can be measured in the laboratory if a good specimen of coal can be obtained in the field. An example is presented here.

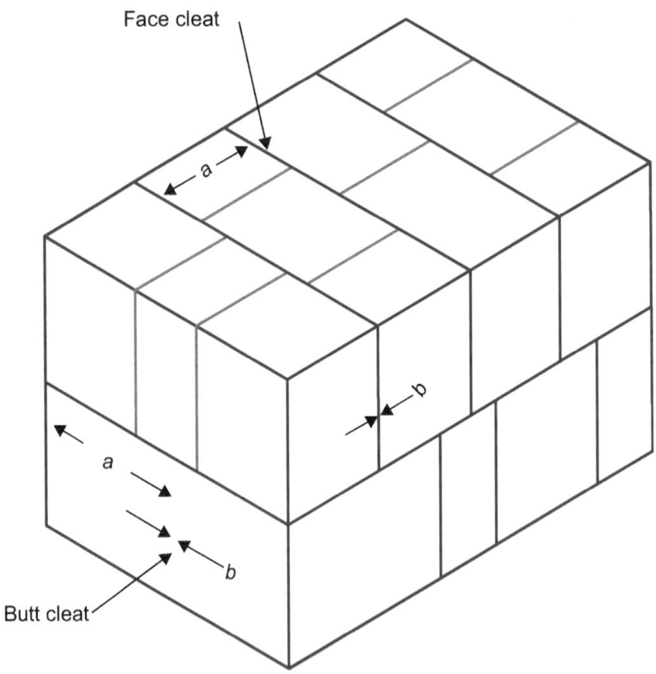

Figure 3.2 Idealized coal block with macrofractures.

$$\text{Assume } a = 1 \text{ mm} = 1000\ \mu\text{m}$$
$$b = 5\ \mu\text{m}$$
$$\text{Then porosity} = \phi = \frac{3\,b}{a} = \frac{3 \times 5}{1000} = 0.015 \quad \text{or } 1.5\% \tag{3.10}$$
$$\text{Permeability} = \frac{b^3}{12\,a} = \frac{5^3 \times 1000}{12 \times (1000)} = 10.42 \text{ md}$$

For better accuracy, both fracture widths (face cleat, b_1, and butt cleat, b_2) should be measured and an average width should be calculated: $b = \frac{b_1 + b_2}{2}$.

No history matching data from the field is yet available to confirm the accuracy of these calculated porosity or permeability values.

3.4.2 Laboratory Methods of Permeability Measurement

The conventional, steady state, technique of measuring permeability used a polished core, typically 1–1.5 inches in diameter and 1–2 inches long. It was confined in a sleeve and pressurized to simulate in situ conditions. Fluid flow along the specimen under a given applied pressure was measured and permeabilities[1] were calculated using Eq. (3.5) or (3.7) for liquid or gas respectively. The values obtained are usually much lower than the actual matrix permeability. For low-permeability rocks, such as coal and shale, it also took a long time to achieve a steady state.

Unsteady state methods such as GRI (Gas Research Institute) and pressure pulse decay have been used to estimate the permeability of shale samples because they are faster and can measure permeability in the nanodarcy range [3,4].

3.4.2.1 The GRI Technique

This technique is similar to Boyle's Law double cell porosity measurement [5]. An inert gas such as helium is expanded from a reference cell, at pressure P_1, into a second cell containing the crushed shale or coal. Tinni [3] shows the experimental details and calculations, but the technique suffers from a number of drawbacks.

a. The sample cannot be confined to duplicate in situ stress
b. The gas flow may not follow Darcy's law (pressure-dependent flow)
c. The method is highly dependent on the grain size.

[1] It is advisable to take five to six measurements and average the data.

The GRI method typically shows 3–10 times higher permeability than that found by steady state measurements [6].

3.4.2.2 Pulse Decay Technique

The pulse decay technique was first developed by Brace [7] for granite but it can be used to measure the permeability of other rocks. In this technique, a core plug under confining pressure is brought to an equilibrium pressure. Next, a pressure pulse is imposed on the upstream side of the plug, and the pressure decay and the build-up are recorded over time on the upstream and downstream sides, respectively. The change in the pressure pulse with time is then analyzed to estimate the permeability. The natural logarithm of pressure is plotted against time. The slope of the plot is a function of permeability, and a transient Laplace equation is used to calculate permeability [4].

The technique has been further refined but it remains time-consuming and difficult to interpret. The permeability values obtained by this technique are 2–8 times higher than those found by steady state techniques [8].

Further modifications in these techniques are in progress [9], but they are unlikely to be useful for coal deposits because coal is greatly nonhomogenous and sorptive. The only viable data on coal permeability is the effective permeability of the reservoir measured in the field.

3.4.3 Field Measurement of Permeability

Field measurement of coal permeability provides the most accurate data. The techniques used for this purpose can be broadly classified into two categories:

1. Pressure transient tests.
2. History matching of gas production data.

3.4.3.1 Pressure Transient Tests

Natural gas wells have been tested by the following methods:

1. Drill stem testing.
2. Slug testing.
3. Injection fall-off testing. This method has three variations: tank testing, pressure injection without fracking, and mini-frack injection testing.
4. Pressure build-up and drawdown testing.
5. Multi-well interference testing.

Detailed descriptions of these techniques are available in the literature [1,10].

Only the three techniques that are best for coal permeability measurement will be discussed here.

3.4.3.1.1 Mini-Frack Injection Testing

In this test, a small volume (1000–2000 gallons) of 2% KCl water is injected into the coal formation at a low rate of 3–5 bbl/minute. Normal fracking of coal is done at a much higher rate of injection, 30–35 bbl/minute. Bottom hole pressure (BHP) is continuously measured as the mini-frack progresses. The buildup of pressure until the coal minimally fracks is recorded. The injection is stopped as soon as about 1000–2000 gallons have been pumped in. The BHP at this point is immediately recorded. This pressure is called the instantaneous shut-in pressure (ISIP). The sum of the ISIP plus the hydrostatic head, divided by the depth of the borehole, is called the "frack gradient." It can be used to predict permeability by history matching. A typical plot of BHP against time is shown in Fig. 3.3. The rate of pressure decline changes twice before it becomes steady.

To clarify, the decline curve in Fig. 3.4 is plotted as dp/dt against time. The first inflection point is called the fissure opening pressure and the second inflexion point is called the closure pressure. Like the frack-gradient, closure pressure can be used to estimate permeability.

Fig. 3.5 shows a relationship between the permeability and the closure pressure for a US coal field. In general, the higher the closure pressure, the lower the coal permeability.

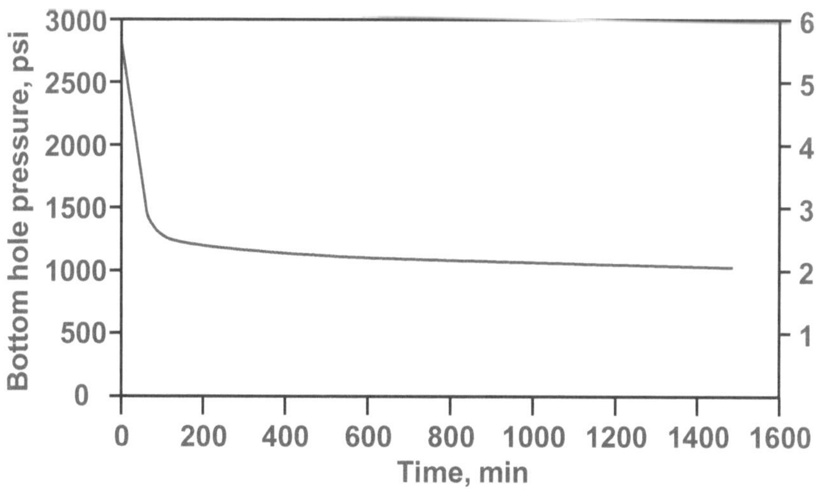

Figure 3.3 A typical mini-frack pressure versus time graph.

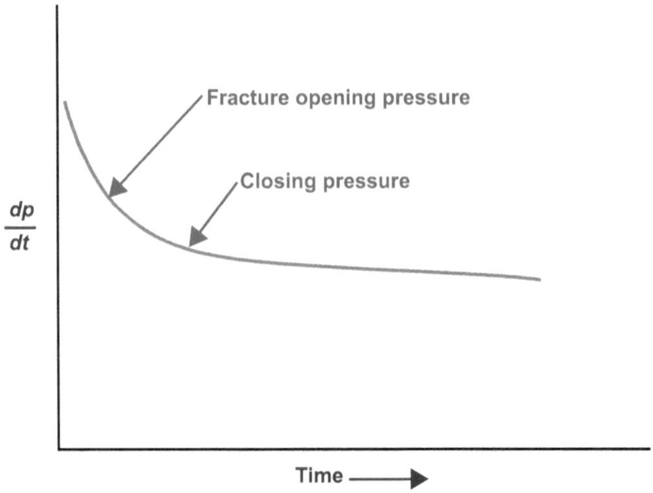

Figure 3.4 Pressure decline rate versus time.

Mathematically,

$$\ln k = a + b\,P_c \tag{3.11}$$

where

k is the permeability in md
a and b are constants for the coal seam
P_c is the closure pressure

Analyzing the data in Fig. 3.5,

$a = 5.1$
$b = -0.001485$

3.4.3.1.2 Calculation of Effective Reservoir Permeability From a Drawdown Curve

It will be shown in a later chapter that

$$P_s^2 = \frac{-\,\mathrm{m}\,P_f^2 \ln t}{2} + c \tag{3.12}$$

This equation applies for an infinite reservoir with laminar flow at large values of dimensionless time. P_s is the BHP and t is time. If we plot P_s^2 against ln t, the slope of the straight line is $\frac{-\,m\,P_f^2}{2}$

The slope is related to permeability by the equation

$$k = \frac{1424\,\mu\,ZTQ}{2h \times \text{slope of the } P_s^2 \text{ plot}} \tag{3.13}$$

A simple way of finding the slope is to plot P_s^2 versus ln t on semi-log paper. The slope will be read in psia^2 per cycle, divided by 2.303.

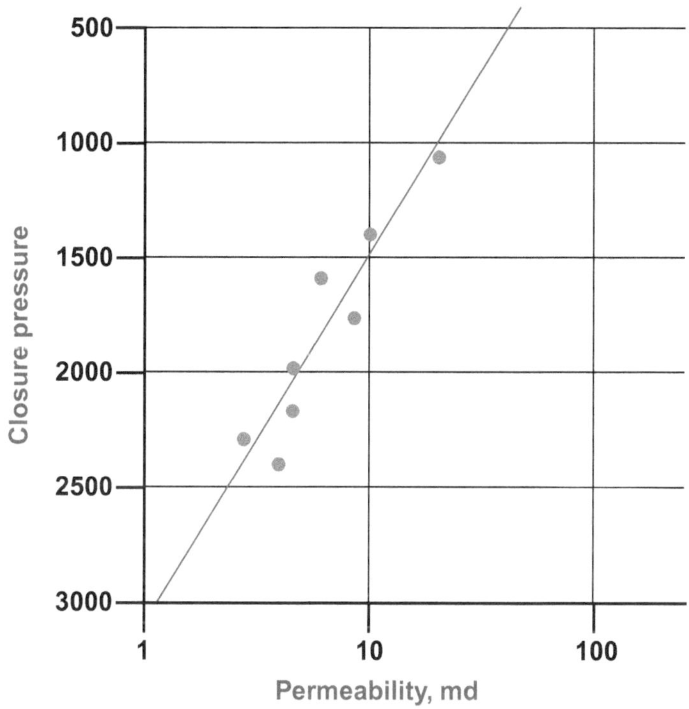

Figure 3.5 Permeability versus closure pressure.

In the following example it is 10,000 squared psi per cycle.

A typical plot for a coal bed methane well is shown in Fig. 3.6.

An example is provided here. The necessary data is listed in Table 3.1.

Using the above data in Eq. (3.13):

$$k = \frac{1424 \times 0.011 \times 0.95 \times 525 \times 275 \times 2.303}{2 \times 5 \times (10,000)} = 49.47 \text{ md} \quad (3.14)$$

3.4.3.1.3 Calculation of Effective Reservoir Permeability From a Build-Up Curve

It will be shown in a later chapter that

$$\frac{P_s^2 - P_f^2}{P_f^2} = \frac{-m_1}{2}(\ln t_{d_1} - \ln t_{d_2})$$

or (3.15)

$$P_s^2 = \frac{-m_1 P_f^2}{2}(\ln(t_f + \Delta t) - \ln \Delta t) + P_f^2$$

Δt = length of time since the shut-in, and

P_s and P_f are in psia.

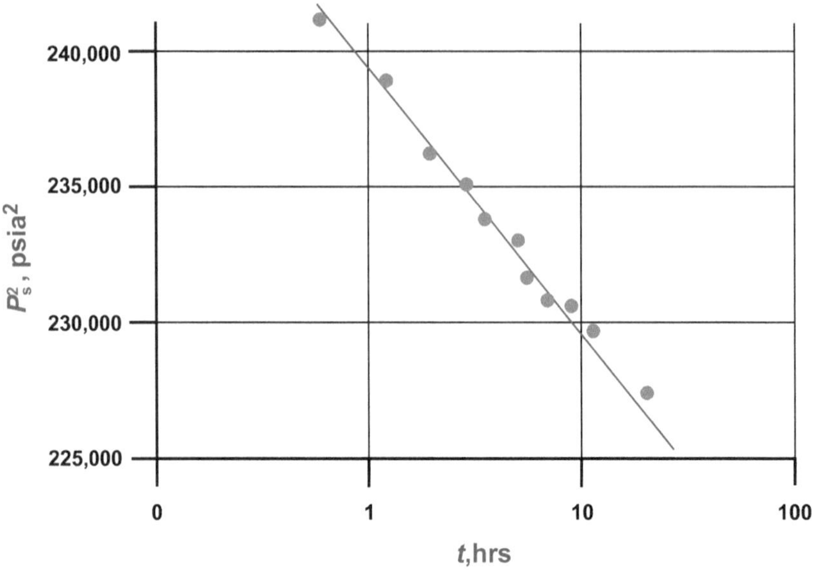

Figure 3.6 BHP^2 versus ln t graph for a drawdown test.

Table 3.1 Reservoir data for permeability calculation

Depth in feet	**2000**
Reservoir temperature, °F	65
Gas gravity	0.57
Coal thickness, ft	5.00
Gas compressibility in reservoir	0.95
Gas viscosity at 500 psi, cp	0.011
Closed reservoir pressure in psi	500
Flow rate on drawdown test, MCFD	250–300
Average, MCFD	275

For a long period of flow, a simpler type of relation between pressure and time is created.

Differentiating the above Eq. (3.15), Eq. (3.16) is obtained:

$$\frac{dP_s^2}{d(\Delta t)} = \frac{-m_1 P_f^2}{2}\left[\frac{1}{t_f + \Delta t} - \frac{1}{\Delta t}\right] \tag{3.16}$$

If t_f is long (the well has been producing for a long time) the term $\frac{1}{t_f + \Delta t}$ will be negligible.

$$\text{Hence, } \frac{dP_s^2(\Delta t)}{d(\Delta t)} = -\frac{m_1 P_f^2}{2}$$

or

$$\frac{dP_s^2}{d(\ln\Delta t)} = -\frac{m\ P_f^2}{2} \tag{3.17}$$

which is the slope of the curve of P_s^2 versus $\ln\left(\frac{t_f + \Delta t}{\Delta t}\right)$

Fig. 3.7 shows a plot of a typical build-up curve. The slope is 9000 psia^2 per cycle.

Hence,

$$k = \frac{1424 \times 0.011 \times 0.95 \times 525 \times 275 \times 2.303}{2 \times 5 \times 9000} = 54.9 \text{ md} \tag{3.18}$$

Field data on coal permeability available in the literature and measured by the author are listed in Table 3.2.

Another measure of net deliverability of a coal seam is obtained by drilling a 1000 ft long horizontal borehole and measuring the specific production of gas from the coal seam. This data can be used to calculate the effective permeability or the effective length of fracture wings in a coal seam, as shown later in this book. Specific production for some coal seams is also shown in Table 3.2.

3.5 FACTORS INFLUENCING THE RESERVOIR PERMEABILITY

Many factors, such as depth, reservoir pressure, ground stress, the Klinkenberg effect, and shrinkage due to gas desorption, impact the effective permeability. Only some of these factors that are known to have a significant impact are discussed here.

3.5.1 Depth of Coal Seams and Ground Stress

The depth of the coal seam correlates well with ground stress. Both tend to reduce coal permeability. As discussed in Chapter 2, Gas Content of Coal and Reserve Estimates, deeper coal seams contain higher amount of gas per ton of coal but the limiting factor for commercial gas production is the coal permeability.

Analyzing the data available in the literature, Fig. 3.8 shows that the relationship between depth and permeability in a coal basin can be expressed as

$$K = K_o\, e^{-\mathrm{aD}} \tag{3.19}$$

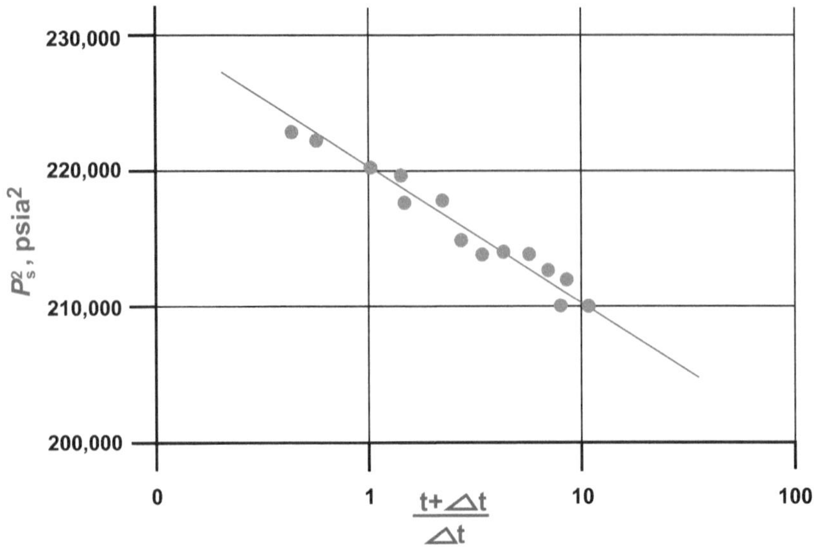

Figure 3.7 Pressure^2 versus $\left(\frac{t_f + \Delta t}{\Delta t}\right)$.

Table 3.2 Effective permeability & related data for US coal fields

Basin/coal seam	Depth (ft)	Closure pressure (psi)	Effective permeability (md)	Specific production (MCFD/100 ft)
Northern Appalachia Pittsburgh seam	500–1000	...	10–50	15–25
Central Appalachia Pocahontas 3 seam	1500–2500	1300–2000	1–20	7–8
[a]Southern Appalachia Mary Lee-Blue Creek	1500–2500	1300–2500	10–25	5–7
Southern Appalachia Oak Grove field	1000		10–50	...
[a]San Juan Basin	2000–3000		1.5–8.8	N/A
European coal fields	3000 +		Less than 1	N/A

[a]Adapted from Rogers [1].

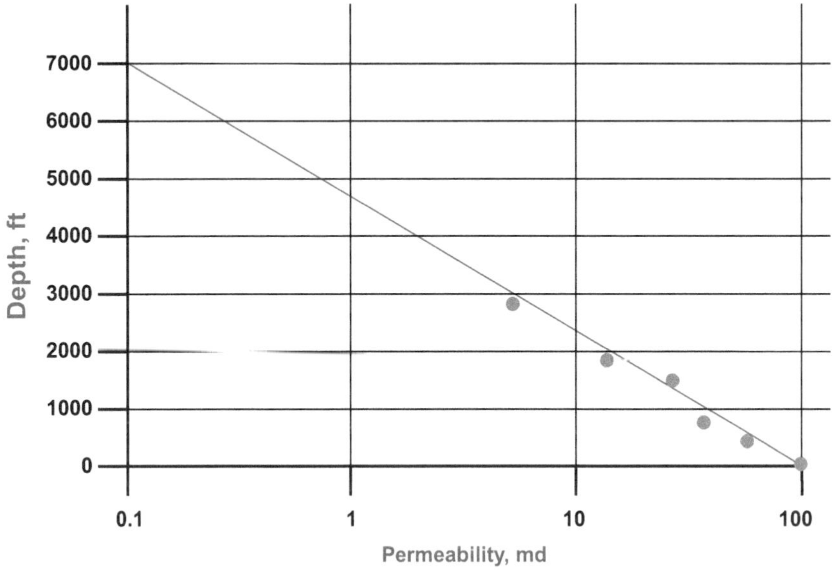

Figure 3.8 Permeability versus depth.

where

K_o is the permeability at 100 ft depth

a is a constant for a coal basin, with a value of 700—1000 ft.

D is the depth in feet

Thus the permeability of coal in a basin at 2000 ft deep is $K = 100\, e^{\frac{-2000}{700}} = 5.7$ md. If $a = 1000$ ft; $K = 13.5$ md.

Similarly, as the horizontal stress increased with depth, the permeability can be expressed as $K = K_o\, e^{-b\sigma}$[11], where σ is the net major, horizontal principal stress.

At 2000 ft in a Virginia coal basin, the major horizontal stress σH is 3400 psi, and the minor stress, σh, is 1700 psi. Vertical stress is usually equal to $1 \cdot 1$ D/psi, where D is the depth in feet. $\sigma = \sigma_H - \sigma_0$ where σ_0 is the pore pressure (gas pressure) of the coal seam.

The value of 'b' is on the order of 8×10^{-3}.

3.5.2 Coal Seam Temperature

Increasing temperature is likely to increase the permeability. Mathematically,

$$K = K_o(1 + T)^n \tag{3.20}$$

All these parameters have to be experimentally determined [11].

This hints at the possibility of commercial gas production from deep, low-permeability coal reservoirs by heating the coal seam. This subject will be discussed in detail in a subsequent chapter.

3.5.3 Effect of Reduction in Reservoir Pressure/Shrinkage of the Coal Matrix

Many researchers have proved that coal swells as it adsorbs gas under pressure. Similarly, as the coal desorbs gas with a fall in pressure, it shrinks, enhancing permeability and the rate of gas production.

Li [12] shows in great detail the relationship between the original permeability and the enhanced permeability owing to reduction in reservoir pressure.

$$K = \frac{K_o}{1 + E_v}\left\{1 + \frac{E_v - E_p}{Q_o}\right\}^3 \tag{3.21}$$

where

K_o is the original permeability of coal
E_v is the volumetric strain due to stress change
E_p is the volumetric strain due to desorption/shrinkage
E_p is expressed as:

$$E_p = \frac{a\,K_c R\,T}{V_o} \ln\left(1 + b\,P\right) \tag{3.22}$$

where

a, b, P are from the Langmuir equation (in chapter: Gas Content of Coal and Reserve Estimates);
K_c is a constant
V_o is the volume of a mole of gas at standard temperature and pressure
R is the gas constant, and
T is the absolute temperature

3.5.4 The Klinkenberg Effect

The permeability of porous rock to various gases appears to increase with the reciprocal of pressure in the reservoir. Thus as the reservoir pressure declines, an increase in gas production over what is predicted by Darcy's law is observed. This is known as the Klinkenberg effect.

Mathematically,

$$K = K_o\left(1 + \frac{b}{P}\right) \tag{3.23}$$

where

K is the apparent permeability

K_o is the original permeability at original reservoir pressure

b is the slippage factor

P is the mean pressure

Combining the Klinkenberg effect with reservoir shrinkage enhancing permeability further, it is normal to expect a coal seam gas well to continue to produce at commercial rates at much lower pressures compared to a conventional well in porous sandstone.

For deeper coal, permeability can be enhanced by either radio frequency heating or in situ combustion. These production enhancement techniques will be discussed later in this book.

REFERENCES

[1] Rogers R, et al. Coal bed methane: principles and practices. Oktibbeha Publishing LLC; 2007. p. 504.

[2] Robertson EP, Christensen RL. A permeability model for coal and other fractured, sorptive-elastic media, U.S. Department of Energy, INL/CON-06-11830, 2006. p. 26.

[3] Tinni AF. Shale permeability measurement on plugs and crushed samples, Paper SPE 162235 presented at Unconventional Resources Conference, Calgary, Alberta, Canada; 2012.

[4] Jones S. A technique for faster pulse decay permeability measurements in tight rock. SPEFE 1997;19–25.

[5] American Petroleum Institute. Recommended practices for core analysis, 40, Washington, D.C., 1998.

[6] Eagerman. A fast and direct method of permeability measurement on drill cuttings, SPE Paper 77563, 2005.

[7] Brace WW. Permeability of granite under high pressure. J Geophys Res 1968;73–2225.

[8] Rushing JN. Klinkenberg—connected permeability measurements in tight gas sand: steady-state versus unsteady-state techniques. SPE Annual Technical Conference, Houston, Texas, SPE 89867, 2004.

[9] Zaminian M. New experimental approach to measure petrological properties of organic rich shales, PhD Thesis. Morgantown, WV: West Virginia University; 2015. p. 114.

[10] Katz DL, et al. Handbook of natural gas engineering. New York, NY: McGraw-Hill Book Company; 1958.

[11] Gui X, Meng X. Application of coal gas permeability factor and gas transportation. Chin J Min Saf Eng 2008.

[12] Li X, et al. Coal Operators Conference In: Aziz N, editor. Analysis and research on the influencing factors of coal reservoir permeability. University of Wollongong; 2008. p. 197–201.

CHAPTER 4

Diffusion of Gases From Coal

Contents

Abstract

The complex flow of gases from the coal matrix is basically controlled by (1) diffusion of gas from coal following Fick's law and (2) Laminar flow of gases through the fractures in coal following the Darcy's law or pressure-dependent flow. The slower of the two processes decides the net flow because they work in series. Diffusion process is fully explained. Methane is stored on coal (molecule) surfaces adsorbed in a monolayer. A new parameter that controls the rate of diffusional flow is defined as $\left(\frac{D}{a^2}\right)$ second^{-1} where D is the coefficient of diffusivity and a is the hypothetical coal particle radius where methane is adsorbed. In mining parlance, the value of $\left(\frac{D}{a^2}\right)$ is not easily realized and hence a new parameter, τ, was created that is sorption time of coal. During this time, the coal will diffuse about 63% of the total gas contained on its surface.

Sorption time, τ, was expressed as $\frac{3.49 \times 10^{-2}}{\left(\frac{D}{a^2}\right)}$ and the value of D for various coal seams was calculated from measured values of τ. The sorption time has a great impact on the recovery percentage of in-situ gas reserve for a given time of production. A simpler technique for determining τ was derived and shown in Eq. (4.14). Plotting the gas desorbed expressed as a fraction of the Langmuir volume against $t^{1/2}$ or $t^{1/3}$ will yield a straight line. The slope of the line is equal to $\left(\frac{1}{\tau}\right)^{1/2}$ or $\left(\frac{1}{\tau}\right)^{1/3}$ respectively. Finally, the influence of pressure and temperature on diffusivity was discussed. Diffusivity increases as pressure goes down and temperature goes up. Thus heating the coal seam can enhance gas production.

Advanced Reservoir and Production Engineering for Coal Bed Methane.
DOI: http://dx.doi.org/10.1016/B978-0-12-803095-0.00004-1

Methane is held in adsorption on the surface of coal particles in a monolayer.

The flow of gases adsorbed on the coal matrix surfaces starts as soon as the confining pressure is reduced. The process goes through the following steps:

- Diffusion of gas from coal following Fick's law, i.e., concentration-dependent flow.
- Laminar flow of gases through the fractures in coal matrix. This follows Darcy's law, i.e., pressure-dependent flow. It is controlled by permeability.
- Turbulent gas flow in horizontal boreholes and vertical wells. This is controlled by the pressure gradient and by borehole/pipeline characteristics.

The net flow of gases is controlled by the first two factors: rate of diffusion and permeability-controlled flow. The sizes of horizontal boreholes and casings are designed to be so large that they do not impede the gas flow. Fig. 4.1 illustrates the flow sequence [1].

The two processes work in series, and the one with a lower rate will control the net flow. Thus in a shallow reservoir with high permeability and a low diffusion coefficient, the diffusivity determines the flow rate. In a deeper coal of high rank, the diffusivity is one or two orders of magnitude higher, and the permeability is lower. Hence, permeability determines the flow rate. It is, therefore, important to analyze the diffusion of gas from coal.

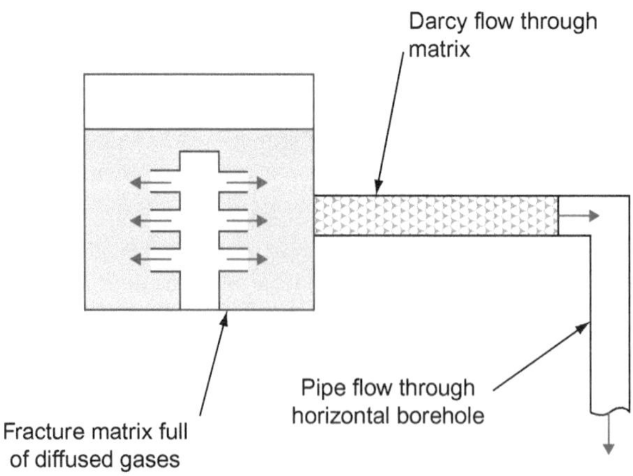

Figure 4.1 A model of methane flow in coalbeds.

The interplay of permeability and diffusivity will also decide the spacing of vertical gas wells and the distance between two horizontal wells drilled from the surface for optimum cost and gas production.

4.1 THE DIFFUSION PROCESS

When a gas composed of molecule A (methane) comes in contact with gas composed of molecule B (air), the contact will cause diffusion of A into B and B into A. The process tends to produce a mixture of a uniform composition. Many enhanced methane production techniques from coal use a gas driver, such as carbon dioxide (CO_2) or helium (He), to increase production. Gas-to-gas diffusion is important part to study to predict the diffusion process. Similarly when gas is passed through a porous medium wet with liquid (oil), the rate of attaining equilibrium between the gas and liquid phases depends on the diffusion process.

Fig. 4.2 shows a simple diffusion process. Container A has CO_2 at 100% concentration. Container B has 99% methane with 1% carbon dioxide. If the two vessels are connected by a conduit 1 cm × 1 cm and 1 cm long, CO_2 will try to go into container B and likewise, methane will try to go into container A.

Assuming the containers are large in relation to the diffusion rate, the process is expressed mathematically as

$$\frac{dc}{dt} = -DA\frac{dc}{dx} \tag{4.1}$$

where

c = number of molecules diffusing
t = time
D = diffusivity coefficient
A = area
$\frac{dc}{dx}$ = concentration gradient

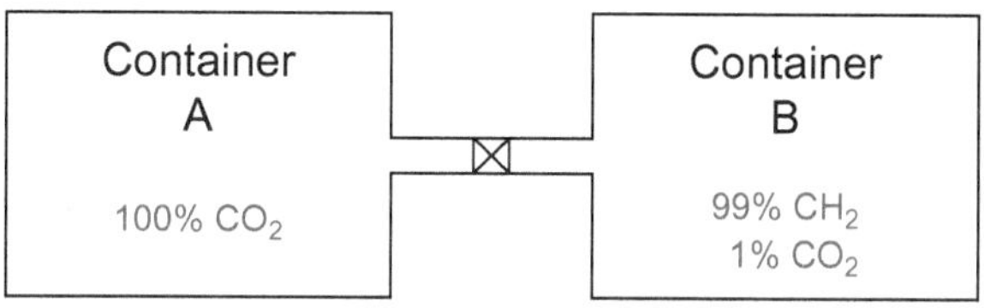

Figure 4.2 Diffusion process for gases.

An example: For the given data in Fig. 4.1 at 0c and D, given the diffusion coefficient of CO_2 in methane = 0.147 cm/second^2, compute diffusion of CO_2 into methane through the conduit 1 cm × 1 cm and 1 cm long.

$$\frac{dc}{dt} = -\left(0.147 \frac{\text{cm}^2}{\text{second}}\right)\left(\frac{1 \times 1\ \text{cm}^2}{1 \quad \text{cm}}\right)\left(\frac{0.99}{22,414} \times \frac{273}{273}\right)\frac{g\text{-mole}}{\text{cm}^3}$$

$$= 6.5 \times 10^{-6}\ g\text{-mole/second}$$

Fig. 4.3 shows a tiny sphere of coal with methane adsorbed on the surface in a monolayer.

Assuming that the flow is radial, the turbulent diffusion equation for a constant diffusion coefficient takes the form

$$\frac{dc}{dt} = D\left(\frac{d^2c}{dr^2} + \frac{2}{r} \cdot \frac{dc}{dr}\right) \tag{4.2}$$

The diffusion process is non-steady.

c is the gas concentration

a is the coal particle radius

t is time

D is the coefficient of diffusion.

The following are the initial and boundary conditions:

$$c_r = 0, \quad r = 0, \quad t > 0 \tag{4.3}$$

$$c_r = ac_0, \quad r = a, \quad t > 0 \tag{4.4}$$

$$c_r = r f(r), \quad t = 0, \quad 0 < r < a \tag{4.5}$$

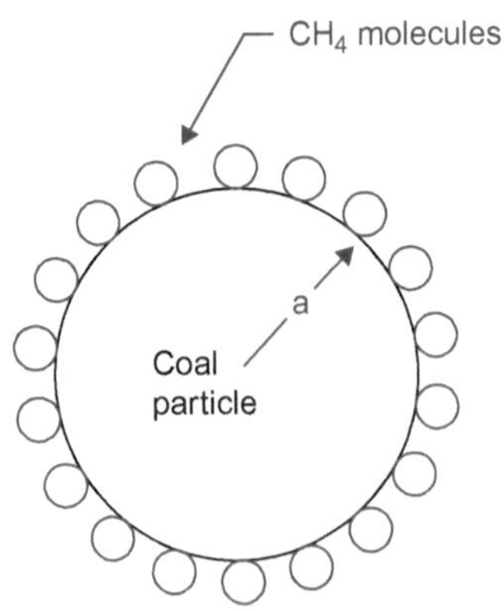

Figure 4.3 Monolayer adsorption of methane on a coal particle.

The solution of Eq. (4.2) for the total amount of gas leaving the sphere is given by Crank [2] as

$$\frac{M_t}{M_\infty} = 1 - \frac{6}{\pi^2}\sum_{n=1}^{\infty}\frac{1}{n^2}\exp\cdot\left(-\frac{Dn^2\pi^2 t}{a^2}\right) \tag{4.6}$$

where M_t is the amount of gas desorbed in time t and M_∞ is the Langmuir volume.

The corresponding solution for a short time is

$$\frac{M_t}{M_\infty} = 6\left(\frac{Dt}{a^2}\right)^{\frac{1}{2}}\left\{\pi^{\ \frac{1}{2}} + 2\sum_{n=1}^{\infty}\operatorname{ierfc}\frac{na}{\sqrt{Dt}}\right\} - \frac{3Dt}{a^2} \tag{4.7}$$

Discarding the term $2\sum_{n=1}^{\infty}\operatorname{ierfc}\frac{na}{\sqrt{Dt}}$ because it is small in value,

$$\frac{M_t}{M_\infty} = 6\left(\frac{Dt}{a^2\pi}\right)^{\frac{1}{2}} - \frac{3Dt}{a^2} \tag{4.8}$$

$$\text{or}\quad \frac{M_t}{M_\infty} = \frac{6}{a}\sqrt{\frac{Dt}{\pi}} - \frac{3Dt}{a^2} \tag{4.9}$$

Since D is on the order of 10^{-10}, we can discard the $\frac{3Dt}{a^2}$ term.

Note: $\sqrt{D} \gg D$ when $D \approx 10^{-10}$

Eq. (4.9) can be rewritten as

$$\frac{M_t}{M_\infty} = \frac{6}{\sqrt{\pi}}\left(\frac{D}{a^2}\right)^{\frac{1}{2}} t^{\frac{1}{2}} \tag{4.10}$$

Since we do not know the actual size of a for coal, which depends on the rank, depth, and friability of the coal, it is better to express diffusivity as $\frac{D}{a^2}$. It has a unit of second^{-1}.

If we plot $\frac{M_t}{M_\infty}$ as a function of $t^{\frac{1}{2}}$ the gradient of the straight line would be equal to $\frac{6}{\sqrt{\pi}}\left(\frac{D}{a^2}\right)^{\frac{1}{2}}$, from which $\frac{D}{a^2}$ can be calculated.

The time taken for a piece of coal to desorb $\left(1 - \frac{1}{e}\right)\%$ or 63% of gas is called τ or "sorption time." This expresses the rate of desorption in mining parlance better than the absolute value of D or $\left(\frac{D}{a^2}\right)$.

Modifying Eq. (4.10), we can write

$$\frac{M_t}{M_\infty} = \left(1 - \frac{1}{e}\right) = \frac{6}{\sqrt{\pi}}\left(\frac{D}{a^2}\right)^{\frac{1}{2}} \tau^{\frac{1}{2}} \tag{4.11}$$

Rearranging and squaring both sides

$$\tau = \frac{\pi\left(1-\frac{1}{e}\right)^2}{36} \Big/ \left(\frac{D}{a^2}\right) \quad \text{or} \quad \tau = \frac{3.49 \times 10^{-2}}{(D/a^2)} \tag{4.12}$$

A typical value of (D/a^2) is 10^{-8} second^{-1} for methane in coal. Hence, the sorption time = 40.4 days.

Table 4.1 shows the values of $\frac{D}{a^2}$ for some known sorption times of various US coal seams.

For comparison, the diffusion coefficient for a gas-to-gas transfer is given in Table 4.2.

Table 4.1 Sorption time and diffusivity for some US coals

Coal seam	Sorption time (day)	D/a^2 (second^{-1})
Pittsburgh	100–900	4.00×10^{-9} to 4.4×10^{-10}
Pocahontas #3	1–3	4.0×10^{-7} to 1.34×10^{-7}
Mary Lee/Bluecreek (Alabama)	3–5	1.34×10^{-7} to 8×10^{-8}
San Juan Basin Coal	1	$< 4.0 \times 10^{-7}$

Table 4.2 Diffusion coefficient of gases at atmospheric pressure

System	Temperature (°C)	Diffusion coefficient, D (cm^2/second)
Methane in air	0	0.196
Carbon dioxide in air	25	0.164
Carbon dioxide in methane	0	0.147
Hydrogen in air	25	0.410
Hydrogen in methane	0	0.630
Methane in methane	19	0.214

Source: Adapted from Katz DL. Handbook of natural gas engineering. McGraw-Hill Book Company, 1958, p. 100 [3].

4.2 AN EMPIRICAL EQUATION FOR DIFFUSION OR GAS DESORPTION FROM COAL

Airey [4] proposed an empirical equation to determine the diffusion of gases.

$$\frac{M_t}{M_\infty} = 1 - \exp\cdot\left(-\frac{t}{\tau}\right)^n \tag{4.13}$$

where τ is the characteristic time, or sorption time for 63% desorption of gas

$n = 1/2$ for anthracite

$n = 1/3$ for bituminous coal

Expanding Eq. (4.13)

$$\frac{M_t}{M_\infty} = \left(\frac{t}{\tau}\right)^n + \frac{1}{2}\left(\frac{t}{\tau}\right)^{n^2} + \cdots$$

and discarding $\left(\frac{t}{\tau}\right)^{n^2}$ and higher terms,

$$\frac{M_t}{M_\infty} = \left(\frac{t}{\tau}\right)^n = K\cdot t^n, \text{ where } K = \left(\frac{1}{\tau}\right)^n \tag{4.14}$$

Plotting $\frac{M}{M_\infty}$ against $t^{1/2}$ or $t^{1/3}$, the term K can be determined. From K, τ, the sorption time for the particular coal, can be worked out.

4.3 ANOTHER EMPIRICAL RELATIONSHIP FOR τ, THE SORPTION TIME

King and Ertekin [5] have suggested another empirical relationship for τ as

$$\tau = \frac{1}{8\pi}\left(\frac{1}{D/S^2}\right) \simeq \frac{4.0\times 10^{-2}}{D/S^2} \tag{4.15}$$

where S is the spacing between major cleats, the particle diameter is related to S by

$$a = \left(\frac{8\pi}{S^2}\right) \tag{4.16}$$

Substituting for S^2 in Equation 4.15, $\tau = 1.39\times 10^{-3}/Da$. This appears to be dimensionally inconsistent.

4.4 FACTORS THAT INFLUENCE DIFFUSIVITY

The diffusivity coefficient is dependent on a number of factors, but the most important are the following:

- type of gas and porous medium interface
- reservoir pressure
- reservoir temperature
- gas-to-gas interface when binary gases are considered

4.4.1 Diffusion Through Porous Media

In general, the rate of diffusion into a porous medium is a lot less than in an empty space because of the restrictions imposed by the solid matrix. In one study [6] where diffusion of CO_2 into a porous metal plug was studied, the diffusivity was reduced by a factor of 4, that is,

$$\frac{\text{Diffusion coefficient of empty space}}{\text{Diffusion coefficient of porous solid}} = 3.94$$

4.4.2 Impact of Pressure on the Diffusivity Coefficient

In general, the diffusion coefficient is inversely proportional to pressure. This is also an observed fact: gas production rates from coal seams tend to increase as the reservoir pressure goes down. This is because of increased diffusivity as well as increased permeability, as discussed in Chapter 3, Porosity and Permeability of Coal. The abandonment time of coal bed methane wells is thus extended since the wells produce longer than anticipated.

Since the density of the gas is directly proportional to pressure at constant temperature, it is generally observed that the product of diffusivity and gas density remains constant over a large pressure range [7].

4.4.3 Impact of Temperature on the Diffusivity Coefficient

Temperature increase generally increases the diffusivity and therefore the rate of gas emission from coal.

Gilliland [8] shows that D is proportional to $T^{3/2}$ for elastic sphere–type molecules where T is the absolute temperature. The typical gas temperature in coal seams is 333K. If the temperature can be raised by 200°C, i.e., to 573K, the rate of gas emission will increase by a factor of more than two.

$$\left[\left(\frac{573}{333}\right)^{1.5}\right] = 2.21$$

This phenomenon has great application in producing gas from deep reservoirs. Gas production can be considerably increased by enhancing the diffusivity just by heating the coal. This will be discussed in greater detail in Chapter 12.

4.4.4 Effective Diffusivity of a Mixture of Gases

In order to enhance gas production from coal seams, either CO_2 or N_2 or both are injected into the coal seams. The net diffusivity of a mixture of gases is given by the following expression [9]:

$$D\,\text{eff}_A = \frac{1 - Y_A}{Y_B/D_{AB} + Y_C/D_{AC} + Y_D/D_{AD}\cdots} \tag{4.17}$$

where

$D\,\text{eff}_A$ is the effective average diffusion coefficient for A in a complex mixture

D_{AB} is the diffusion coefficient of A in System AB

D_{AC} is the diffusion coefficient of A in System AC

Y is the mole fractions and

the A, B, C, D subscripts refer to different gases.

REFERENCES

[1] Thakur PC. Methane flow in the Pittsburgh coal seam, USA. In: Howes MS, Jones MJ, editors. The 3rd International Mine Ventilation Congress. Harrogate, England; 1984. p. 177–82.
[2] Crank J. The mathematics of diffusion. Oxford, UK: Clarendon Press; 1975. p. 91.
[3] Katz DL. Handbook of natural gas engineering. New York, NY: McGraw-Hill Book Company; 1958. p. 100.
[4] Airey EM. Gas emission from broken coal: an experimental and theoretical investigation. Int J Rock Mech Min Sci 1968;5:475.
[5] King GR, Ertekin TM. A survey of Mathematical models related to Methane Production from Coal Seams, Part 1: Empirical and Equilibrium Sorption Models: Proceedings of CBM Symposium, Alabama, United States, 1989, p. 37–55.
[6] O'Hern HA, Martin JJ. Diffusion of carbon dioxide at elevated pressures. Ind Eng Chem 1955;47:2081.
[7] Chou CH, Martin JJ. Diffusion of $C^{14}O_2$ into a mixture of $C^{12}O_2$-H_2 and $C^{12}O_2$-C_3H_8. Ind Eng Chem 1957;49:758.
[8] Gilliland ER. Diffusion coefficients in gaseous systems. Ind Eng Chem 1934;26:681.
[9] Wilke CR. Diffusional properties of multi-component gases. Chem Eng Prog 1950;46:95–104.

CHAPTER 5

Pore Pressure and Stress Field in Coal Reservoirs

Contents

Abstract

Every single point in a CBM reservoir is exposed to four different stresses, namely, the reservoir (pore) pressure, the vertical stress, the major horizontal stress, and the minor horizontal stress. The magnitude of these stresses has a great impact on the selection of the production technique and its success or failure. The pore pressure in most coal seams in the world is about 70% of the hydrostatic head or 0.33 psi/ft of depth. There are some exceptions where the coal seam has a reservoir pressure that is 1−1.2 times the hydrostatic head. Such fields are highly productive. The vertical stress is typically 1.1 × depth, where depth is in feet and pressure is in psi. Horizontal stresses are best derived from equations created from massive data banks collected around the world by research organizations and the world stress map (WSM). Horizontal stress in the rocks containing coal to a depth of 10,000 ft is created by "plate tectonics." The impact of the stress field on production techniques (vertical wells with hydrofracking and horizontal wells drilled from surface) was discussed for major coal fields around the world including Western United States, Eastern United States, Western Europe, Eastern Europe, India, Australia, and South

DOI: http://dx.doi.org/10.1016/B978-0-12-803095-0.00005-3

Africa. Estimation of elastic modulus and poisson ratio of coal seams by sonic logging is discussed. Finally, derivation of bulk modulus, shear modulus, and relationship between various elastic properties is presented. Elastic modulus and shear modulus are needed to design a successful hydrofracturing job.

All underground coal seams have a gas pressure that keeps the gases adsorbed in coal. This is generally called "pore pressure" and designated by σ_0. Similarly, every point in a coal seam has a stress field made up of three stresses:

1. vertical stress, σ_v,
2. major horizontal stress, σ_H, and
3. minor horizontal stress, σ_h.

The stress field and pore pressure, among other reservoir properties, have a major influence on the techniques used for gas production.

As discussed later, horizontal boreholes (with or without hydrofracking) and vertical wells with hydrofracking are two main production techniques. If σ_o is high (typically higher than 500 psi) it is not possible to drill horizontal boreholes in coal. The coal matrix swells/sloughs and seizes the drill steel making it difficult to drill farther. The reservoir pressure must be reduced to less than 200 psi.

For successful horizontal drilling, pore pressure is reduced by vertical drilling and hydrofracking. A good example is the Pocahontas #3 coal seam in Virginia, where the reservoir pressure is 600 psi at a depth of 2000 ft. Horizontal drilling in coal was only feasible when the coal seam was vertically drilled and hydrofracked to lower the pore pressure to approximately 200 psi.

The stress field also determines whether vertical well hydrofracking will be successful. For the well to be productive, the hydrofrack must be vertical—that is, the entire height of the coal seam should be fractured. This is only possible if $\sigma_H > \sigma_v > \sigma_h$, because the plane of a fracture is always perpendicular to the least stress and parallel to the σ_H. In a shallow reservoir, usually $\sigma_H > \sigma_h > \sigma_v$ and hence the fracture plane is horizontal and not very productive. At a depth of 2000 ft or so, $\sigma_H > \sigma_v > \sigma_h$, and hence a good vertical fracture follows with high gas production. The subject will be fully discussed later in the book.

5.1 THE PORE (RESERVOIR) PRESSURE

The reservoir gas pressure appears to be primarily a factor of depth of burial and the rank of coal. The greater the depth of the coal seam and

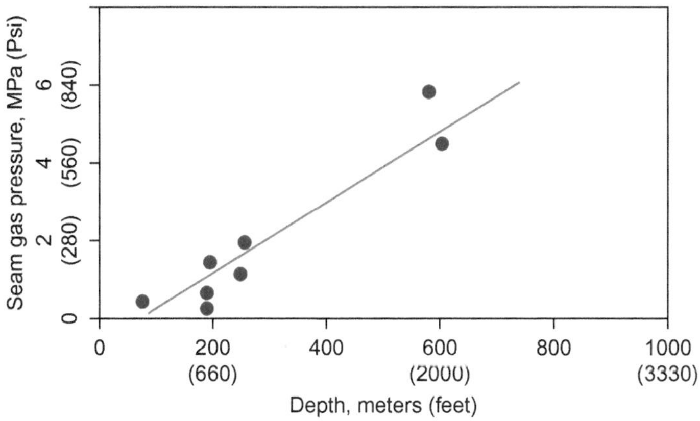

Figure 5.1 Reservoir pressure versus depth.

the higher the rank of coal, the higher the pore pressure. However, there are a few exceptions.

Actual measured pore pressures at various depths from the United States, Canada, Australia and South Africa are plotted in Fig. 5.1 [1].

A linear relationship with depth appears to exist, with a gradient of 0.33 psi/ft. For comparison, the hydrostatic head has a gradient of 0.434 psi/ft. Numerous readings of reservoir pressures in German coal seams showed that pressure is also highly correlated with the rank of coal. The maximum pressure observed in anthracite seams was 700 psi, while that in steam coal was only 250 psi [2]. Some coal seams display a higher pressure gradient than the hydrostatic head gradient of 0.454 psi/ft. Such coal seams are over-pressurized and highly productive. The Fairway region in the San Juan basin of the United States is a good example. Many vertical wells completed in thick, over-pressurized coal seams have had a production of 2–10 MMCFD. The coal seam thickness is 40–60 ft. Similarly, there are a few coal seams that are seriously under-pressurized and are poor producers.

5.1.1 Measurement of Reservoir Pressure

The simplest and perhaps the most reliable technique is to use a pressure gauge, such as an RPG gauge from Halliburton Services.

A vertical well is drilled into the coal seam and extended into the floor for 100–200 ft. A 4.5 inch casing is set in the well just above the coal seam using a formation packer shoe and cemented to the top. The coal seam is hydrojetted with high pressure water at about 3000 psi. Next, an RPG gauge is lowered into the coal seam and a packer is set just above the coal seam. The well is kept shut for 72–96 hours. The gauge shows the pressure build-up on a graph paper. The asymptotic pressure on the graph is the reservoir pressure of the coal seam. Depending on depth, most of the world's coal seams have pressures in the range of 100–800 psi.

5.2 THE VERTICAL PRESSURE, σ_V

It is a commonly accepted fact that

$$\sigma_v = 1.1\ D\,\text{psi} \tag{5.1}$$

where

D is the depth in feet.

This is valid at least to a depth of 10,000 ft. Most coal seams occur above this depth.

5.3 HORIZONTAL (LATERAL) STRESSES IN COAL

For a long time, it was postulated that σ_H and σ_h should be less than σv due to the Poisson effect or roughly equal to σ_v owing to a lithostatic stress state [3]. These hypotheses were soon discarded because they could not explain two peculiar behaviors of horizontal stresses:

1. Why does the horizontal stress often exceed the vertical stress (σ_v) in magnitude?
2. Why are the horizontal stresses so anisotropic? (i.e., $\sigma_H >> \sigma_h$)

Research in the 1970s showed that horizontal stresses are created by the movement of tectonic plates. Plate tectonics state that the crust of the earth consists of a number of continental plates that are sliding over the softer mantle rocks of the earth. The relative movement of plates creates large stresses that are evident in both deep and shallow coal seams [4].

These horizontal stresses can be often measured directly down to the mineable depth of coal seams. Over-coring is the most common technique for measuring stress in underground coal mines. In the United States, most measurements have been made using the US Bureau of Mines biaxial deformation gauge [5]. Internationally, the triaxial ANZI and CSIRO HI cells have been used extensively [6].

Interpretation of data obtained by the over-coring technique requires a good value of the rock/coal's elastic modulus, E.

5.3.1 Estimation of Horizontal Stresses

Christopher and Gadde [4] have done an extensive collection of 565 data points (373 from coal plus 192 from rock) on horizontal stress in coal seams all over the world. They also gathered data from the published World Stress Map (WSM). The range of depth of the coal data was 500–3000 ft. For non-coal, the depth range was 500–8000 ft.

The elastic moduli for coal ranged from 2×10^6 to 6×10^6 psi. The elastic moduli for rocks ranged from 4×10^6 to 9×10^6 psi. Their early conclusions were:

1. Horizontal stresses can exceed the vertical stress by a factor of three or more.
2. $\sigma_H = 2.3\ \sigma_v$ in reverse faulting regions.
3. $\sigma_H = 1.6\ \sigma_v$ in strike-slip faulting regions.
4. $\sigma_H < \sigma_v$ and $\sigma_h = 0.6\ \sigma_v$ in extension faulting regions.

They researched the subject further and developed regression equations to predict σ_H and σ_h based on a depth gradient and the modulus of elasticity of the rock/coal. The same equation works for both stresses, σ_H and σ_h, but the values of the coefficients are different.

The generalized equation for σ_H or σ_h is

$$\sigma_H \ (\text{or } \sigma_h) = B_0 + B_1 \cdot D + B_2 \cdot E \tag{5.2}$$

where

B_0 is the excess stress in psi
B_1 is the corresponding depth gradient in psi/ft
B_2 is a constant coefficient
D is the depth in feet and
E is the elastic modulus in psi.

The values of B_0, B_1, and B_2 are given in Table 5.1 for various coal fields.

An example:

Calculate the σ_H and σ_h in a coal seam in the Eastern United States at a depth of 2000 ft. The elastic modulus of the coal is 2×10^6 psi.

Using Eq. 5.2 and values of B_0, B_1, and B_2 from Table 5.1:

$$\begin{aligned}\sigma_H &= 369 + 1.34 \times 2000 + 0.30 \times 10^{-3} \times 2 \times 10^6 \\ &= 3649 \text{ psi}\end{aligned}$$

$$\begin{aligned}\sigma_h &= 369 + 0.42 \times 2000 + 0.15 \times 10^{-3} \times 2 \times 10^6 \\ &= 1509 \text{ psi}\end{aligned}$$

Table 5.1 Regression coefficients for Eq. 5.2

Coal field	σ_H (psi)			σ_h (psi)		
	B_0 (psi)	B_1 (psi/ft)	$B_2 \times 10^{-3}$	B_0 (psi)	B_1 (psi/ft)	$B_2 \times 10^{-3}$
Eastern United States	369	1.34	0.30	369	0.42[a]	0.15
Western United States	369	0.66	0.62	369	0.56	0.15
United Kingdom and Germany	− 249	0.55	0.51	− 249	0.42	0.15
India	376	1.29	− 0.04	376	0.42	0.15
Australia (NSW)	− 633	1.78	0.56	—	—	—
Australia (Queensland)	− 210	1.40	0.34	—	—	—
South Africa	866	− 0.03[b]	− 0.01	866	0.42	0.15

[a]Christopher and Gadde [4] show 1.34 which may be a typographical error but field data collected by the author corresponds to 0.42.
[b]The depth gradient for σ_H is most likely incorrect. It cannot be negative. Insufficient data may be the reason.
Source: Adapted from Christopher M, Gadde M. Global trends in coal mine horizontal stress measurements, Proceedings of the 27th International Conference on Ground Control in Mining, Morgantown, West Virginia, United States, 2008, p. 319–331.

These values agree well with measured values of σ_H and σ_h in the Central Appalachian coal fields.

σ_v at this depth is $1.1 \times 2000 = 2200$ psi. This creates an ideal condition for hydrofracking because $\sigma_H > \sigma_v > \sigma_h$. The resulting "frac" will be vertical, traversing the entire coal seam and highly productive. These conclusions were verified in the field and will be discussed in Chapter 8, Hydraulic Fracking of Coal.

5.3.2 The Direction of σ_H

Christopher and Gadde [4] provide a summary of the direction of σ_H in various parts of the world. In the United States, the σ_H direction is in the northwest quadrant for the western basin, north-northeast for the central United States and northeast for eastern coal fields. In Western Europe it is typically in the northwest quadrant. In Australia, the direction is quite variable over the continent but the eastern coal fields show the σ_H direction in the northeast quadrant.

Usually the major cleat (face cleat) direction in all coal seams is parallel to σ_H. As discussed earlier, the direction of σ_H is taken into consideration when drilling a horizontal borehole in a coal seam. It should preferably be drilled orthogonal to σ_H for maximum production. If this borehole is hydrofracked to enhance gas production from a deep coal seam, the fracture will run parallel to σ_H, enhancing gas production considerably.

5.4 IMPACT OF THE STRESS FIELD ON PRODUCTION TECHNIQUES

Because of great variations in parameters, a basin-by-basin discussion will be presented to further explain the importance of the four stresses in coal for gas production.

5.4.1 Western US Coal

Vertical wells produce well to a depth of about 3000 ft. Beyond that depth permeability declines fast and horizontal drilling from the surface becomes necessary.

Case 1: Fracking a vertical well at 3000 ft depth; E for coal 3 × 10^6 psi

σ_H is calculated using Eq. 5.2 and data in Table 5.1 to be 4209 psi.

$\sigma_h = 2499$ psi

σ_v from Eq. 5.1 = 3300 psi.

Because $\sigma_H > \sigma_v > \sigma_h$; a vertical fracture will be obtained and good gas production will result. This is confirmed by many highly productive wells in the field. The direction of frack will be generally northwest but there will be some local variations.

Case 2: Horizontal wells at 6000 ft drilled from the surface

Using the same data, the stress field is now as follows:

$\sigma_H = 6189$ psi

$\sigma_h = 4179$ psi

$\sigma_v = 6600$ psi

So $\sigma_v > \sigma_H > \sigma_h$

Hence, hydrofracking the horizontal laterals will result in horizontal fractures in the coal seam. The interval of hydrofracking should be large enough to avoid excessive interference between two adjacent fracs.

There is no field data available to verify the gas production but it is likely to be significantly higher than a single vertical well if two laterals of 5000 foot length are drilled and hydrofracked at 1000 foot intervals.

This production technique has worked very well in the Marcellus shale of the northeastern United States.

5.4.2 Eastern US Coal

The coal deposits in this basin occur only to a depth of about 3000 ft. Only two cases will be considered.

5.4.2.1 Northern Appalachian Basin

Case 1: Shallow deposits to a depth of 1200 ft or less

Field observation show that the hydrofracs are generally horizontal to a depth of 1200 ft; that is $\sigma_H > \sigma_h > \sigma_v$. The data in Table 5.1 confirms that $\sigma_v = \sigma_h$ at a depth of 1204 ft.

Hence hydrofracking of coal seams shallower than 1200 ft will result in a horizontal frac and very little gas production. In an experiment, eight wells were hydrofracked at a depth of 1000 ft but none of them produced measurable quantities of gas.

Shallow deposits are best produced by drilling horizontal wells from surface. The permeability is quite high and hence hydrofracking of horizontal laterals is not necessary.

Case 2: Deeper coal seams

Assume $D = 1500$ ft; $E = 3 \times 10^6$ psi

σ_0 is typically 300 psi

Using data in Table 5.1,

$\sigma_H = 4179$ psi

$\sigma_h = 1449$ psi

$\sigma_v = 1650$ psi

Under this stress field, a vertical hydrofrac will result, yielding good gas production. Thus hydrofracking of vertical wells can be successfully used for coal seams deeper than 1500 ft in this basin.

5.4.2.2 Central and Southern Appalachian Basins

Productive coal seams are generally 2000–2700 ft deep. Coal is brittle and of high rank.

Assume $D = 2500$ ft; $E = 2 \times 10^6$ psi.

σ_0 is typically 500–650 psi.

Using the data in Table 5.1:

$\sigma_H = 4319$ psi

$\sigma_h = 1719$ psi

$\sigma_v = 2750$ psi

Since $\sigma_H > \sigma_v > \sigma_h$, hydrofracking of these coal seams will result in vertical fractures and excellent gas production. Thousands of successful gas wells drilled in these fields were mined out. Our observations confirm the above conclusion.

5.4.3 Western European Coal Basins

There are very few shallow coal seams in the area that are amenable to gas production. Most coal seams are at a depth of 3000–4000 ft.

Assume $D = 3000$ft; $E = 3 \times 10^6$ psi

σ_0 is typically 400–500 psi.

Using the data in Table 5.1:

$\sigma_H = -249 + 0.55 \times 3000 + 0.51 \times 3 \times 10^3 = 2931$ psi

$\sigma_h = 1461$ psi

$\sigma_v = 3300$ psi

Since $\sigma_v > \sigma_H > \sigma_h$, this clearly shows the hydrofracs will be horizontal and hence poor producers. This is confirmed by the data provided in Table 1.4. The direction of frac is NNW or N23°W as confirmed by experiments in German coal mines [7].

With increasing depth there is a decrease in permeability also which adversely impacts gas production.

An exception to the above is the Swansea area of Wales in the United Kingdom, where very gassy coal seams occur at a shallow depth of 2000 ft.

The σ_H and σ_h in this area at 2000 ft depth are 2381 psi and 1041 psi respectively. The σ_v is 2200 psi.

In this case $\sigma_H > \sigma_v > \sigma_h$ and a good hydrofrac will create a vertical frac yielding very good gas production.

Eastern Europe

There is no data on σ_H or σ_h available for this field but assuming Western European data holds good, one can calculate the limiting depth for vertical wells and hydrofracking. At 2329 ft, $\sigma_H = \sigma_v$. At greater depths, where many vertical wells were hydrofracked (refer to chapter 1: Global Reserves of Coal Bed Methane and Prominent Coal Basins), the hydrofracture was horizontal and yielded disappointing gas production.

5.4.4 Indian Coal Fields

There are several coal seams at a depth of 2000 ft that can be hydrofracked for gas production.

Assuming $D = 2000$ ft and $E = 2 \times 10^6$ psi
$\sigma_H = 2876$ psi
$\sigma_h = 1516$ psi
$\sigma_v = 2200$ psi

Here $\sigma_H > \sigma_v > \sigma_h$, hence hydrofracking of these coal seams will result in very good gas production. Because of the depth gradient of 1.29 for σ_H, there is no depth limit for successful hydrofracking of deep coal seams. They are all likely to have a vertical frac resulting in good gas production.

5.4.5 Australian Coal Fields

5.4.5.1 *Sydney Basin*

Coal seams in this basin are deep at 2000 ft and very gassy.

Assume $D = 2000$ ft and $E = 3 \times 10^6$ psi
$\sigma_0 = 500$ psi.

Using data in Table 5.1:
$\sigma_H = 4607$ psi
$\sigma_h = 1073$ psi (B_1 and B_2 assumed as 0.42 and 0.15, respectively)
$\sigma_v = 2200$ psi

Hence $\sigma_H > \sigma_v > \sigma_h$. This will result in a vertical frac with good gas production. With a very high depth gradient for σ_H, there is no depth limit on good hydrofracking.

5.4.5.2 *Bowen Basin*

Coal seams in this basin are typically 1000 ft deep.

Assume $D = 1000$ ft and $E = 2 \times 10^6$ psi
Using the data in Table 5.1:
$\sigma_H = 1530$ psi
$\sigma_h = 660$ psi
$\sigma_v = 1100$ psi

In spite of the shallow depth, the stress field shows $\sigma_H > \sigma_v > \sigma_h$. Hence vertical drilling and fracking should work well, as evidenced by many successful coal bed methane wells in the basin.

5.4.6 South African Coal Fields

Most productive coal seams are shallow, at about 1000 ft, with moderate gas contents. Assume $D = 1000$ ft; $E = 2 \times 10^6$ psi. Data in Table 5.1 cannot be used as explained earlier.

Comparing these coal field with other coal fields, let us assume that the coefficients

for σ_H are $B_0 = 866$ psi, $B_1 = 1.3$ psi/ft, and $B_2 = 0.34$ for σ_h, they are $B_0 = 866$ psi, $B_1 = 0.42$ psi/ft, and $B_2 = 0.15$.

At a depth of 1000 ft:

$\sigma_H = 2846$ psi

$\sigma_h = 1586$ psi

$\sigma_v = 1100$ psi

Hence $\sigma_H > \sigma_h > \sigma_v$. Hydrofracking at this depth will not be successful. However at a depth of 1715 ft $\sigma_v = \sigma_h$. Beyond this depth σ_v will be greater than σ_h and a highly productive vertical fracture will result.

5.5 ESTIMATION OF MODULI OF COAL SEAMS FROM GEOPHYSICAL LOGS

In Eq. 5.2, the only data needed, besides depth, is the elastic modulus of coal to determine the horizontal stresses. For shallow coal seams, a core of the coal seam can be obtained by coring. The elastic modulus can be measured in a laboratory under confined stress that simulates the in situ stress field. Usually such data do not yield a good representation of the actual coal properties. Besides, at great depth, coring is an expensive proposition. Fortunately, gamma ray, density, and sonic logs can be used to derive the value of the elastic modulus reasonably accurately [8].

5.5.1 Sonic Logs

In some logging, ultrasonic frequencies are used in the form of compressional and shear waves. Compressional waves can propagate in solid, liquid, or gas and they exhibit longitudinal particle motion. The shear wave, however, is a transverse wave and its direction of propagation is orthogonal to the direction of particle displacement. Velocity obtained from sonic logs can be used for geophysical evaluation of coal and the surrounding strata. All acoustic velocities are a function of coal/rock density and the elasticity of the rocks. Compressional waves, V_P, are faster and arrive first but shear waves, V_S, make a stronger print even though they are slower.

Detailed coal seam analysis can be obtained with the Formation Micro Images Log (FMI) with its high, 0.2 inch (0.5 cm) vertical resolution. It offers thin coal seam resolution and identification of fractures, cleat type, faults, and in situ stress values. In addition, dipole-type sonic tools can obtain the formation mechanical properties for optimum hydrofracture design. Coal has a higher Poisson ratio and lower Young's modulus than the surrounding shales or sandstone, so coals tend to transfer overburden stress laterally and yield higher fracture gradients. The sonic scanner (SSCAN) is a very good tool. It can not only determine Young's modulus, Poisson ratio, and closure stress gradient but also the in situ stress magnitudes (σ_H and σ_h) and their directions. [8]

The three elastic moduli—Young's modulus, E, bulk modulus, K, and shear modulus, G—and the Poisson ratio υ for coal are mathematically related. They are shown in Table 5.2.

Table 5.3 shows some of the properties of coal types.

Table 5.2 Relations between elastic properties of coal

Elastic constant	Basic equations	Relationships
Young's Modulus	$E = \frac{9K\,V_S^2}{3K + \rho V_S^2}$	$E = 2G(1-\nu)$ $= 3K(1-2\nu)$
Bulk Modulus	$K = \rho V_P^2 - \frac{4}{3V_S^2}$	$\frac{E}{3(1-2\nu)}$
Shear Modulus	$G = \rho V_S^2$	$\frac{E}{2+2\nu}$
Poisson Ratio	$\upsilon = \frac{1}{2}\frac{\left(\frac{V_P^2}{V_S^2}\right) - 2}{\left(\frac{V_P^2}{V_S^2}\right) - 1}$	$\frac{3K - E}{6K}$

where
ρ = density of coal
V_p = compressional wave velocity, μs/ft
V_s = shear wave velocity, μs/ft
υ = Poisson ratio of coal

Table 5.3 Coal rank properties

Coal rank	Density (gm/cm^3)	Sonic Velocity (μs/ft)	CNL Porosity
Anthracite	1.47	105	37
Bituminous coal	1.24–1.34	120	60
Lignite	1.2	160	52

An example:

Assume a bituminous coal has a density of 1.3 (81.1 lb/ft^3) and the sonic velocity, V_S in it was measured at 120 μs (or μs/ft). Calculate the Poisson ratio and all three moduli of elasticity.

$$\text{Given, } V_P = 120\ \mu s;\ V_S \text{ is typically } 0.5\ V_P \text{ in coal} = 60\ \mu s$$

1. Calculate the shear modulus, G

$$G = 81.1 \times (120)^2 = 1.168 \times 10^6 \text{ psi}$$

2. Calculate the Poisson ratio, υ

$$\upsilon = \frac{1}{2} \cdot \frac{4-2}{4-1} = \frac{1}{3} = 0.33$$

3. Calculate $E = 2G(1 - \upsilon)$

$$= 2 \times 1.168 \times 0.67 = 1.565 \times 10^6 \text{ psi}$$

4. Calculate K, the bulk modulus

$$K = \frac{1.565 \times 10^6}{3(1 - 0.66)} = 1.534 \times 10^6 \text{ psi}$$

REFERENCES

[1] Thakur PC. How to plan for methane control in underground coal mines. Min Eng 1977;41–5.

[2] Muche G. Methane desorption within the area of influence of workings, 1st International Mine Ventilation Congress, Johannesburg, S. Africa, 1975.

[3] McGarr. On the state of lithospheric stress in the absence of applied techtonic forces. J Geophys Res 1988;93(B 11):13609–17.

[4] Christopher M, Gadde M. Global trends in coal mine horizontal stress measurements, Proceedings of the 27th International Conference on Ground Control in Mining, Morgantown, West Virginia, United States, 2008, p. 319–331.

[5] Bickel DL. Rock stress determination from over coring – an overview. USBM Bull 1993;694:146.

[6] Mills K. In situ stress measurements using the ANZI stress cell, Proceedings of International Symposium on Rock Stress, Kumamoto, Japan, 1997, p. 149–154.

[7] Mueller W. The stress state in Ruhr Basin, Proceedings of ISRM 7th International Conference on Rock Mechanics, Auchen, Germany, 1991.

[8] Sulton T. Wireline logs for coal bed evaluation. In: Thakur PC, et al., editors. Coal bed methane: from prospect to pipeline. Elsevier Inc; 2014. p. 93–100.

CHAPTER 6

Fluid Flow in CBM Reservoirs

Contents

Abstract

When a gas well is drilled and stimulated in a coal seam, the gas flow goes through three regimes of flow: (1) unsteady flow, which is time-dependent, (2) steady flow, and (3) production decline, also time-dependent. Estimation of gas and water flow under each regime is essential for commercial marketing contracts. First, the unsteady state flow is mathematically modeled and solved to predict pressure decline in a reservoir, where the well is flowed at a constant pressure. Pseudo-porosity for coal is defined and shown to be about 0.55 for most coal. Next, the pressure decline in a coal seam where the gas well is produced at a constant flow rate is calculated. An equation for the cumulative gas flow when the well is produced at a constant pressure from an infinite reservoir is calculated. Then the mathematical derivation of permeability from (1) draw-down test and (2) build-up test are done to supplement the calculation of permeability in Chapter 3, Porosity and

Advanced Reservoir and Production Engineering for Coal Bed Methane.
DOI: http://dx.doi.org/10.1016/B978-0-12-803095-0.00006-5

Permeability of Coal. In the steady-state flow regime, the gas production as well as water production is calculated theoretically. Production increases by hydrofracking a single coal seam and multiple coal seams in a single well are calculated. Finally, six models of production decline are presented. They are: (1) exponential decline, (2) harmonic decline, (3) hyperbolic decline, (4) power law exponential decline, (5) stretched exponential decline, and (6) power law decline. The last was proposed by the author, and three cases of gas production are modeled to confirm the accuracy of the model.

Almost all coal seams are saturated with gas and water. While the gas is mostly methane and carbon dioxide, the water may occasionally contain some soluble impurities, such as sodium chloride up to 3% by weight. When a horizontal borehole or a vertical well is drilled into a coal seam, it produces both gas and water. The water comes out first and inhibits the gas flow. In a few days to a few months, the water flow is minimized and the gas flow peaks. It maintains a steady rate for a short period, after which the gas flow declines.

Fig. 6.1 shows the three phases of gas flow from a vertical well in coal.

The fluid flow from coal reservoirs, therefore, will be covered in three phases:

1. Unsteady state flow
2. Steady-state flow
3. Declining flow or production decline

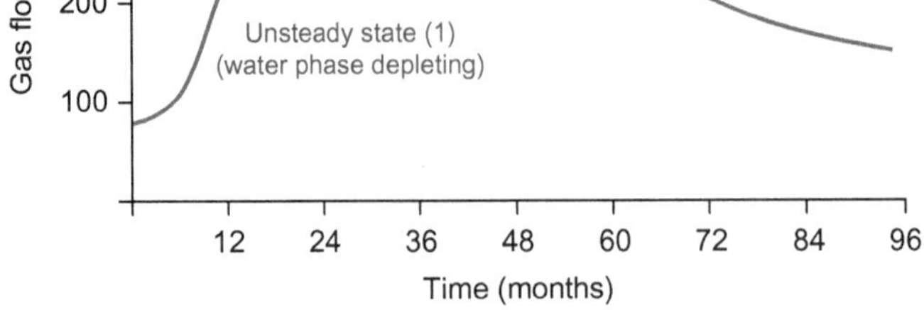

Figure 6.1 Gas flow from a vertical well in a coal seam versus time.

The flow of fluids in coal reservoirs is a vast subject depending on the following variables:

1. Liquid or gas
2. Laminar or turbulent flow
3. Linear or radial flow
4. Steady state or unsteady state flow
5. Finite or infinite reservoir

Only a limited number of cases that are most pertinent to gas production from coal will be discussed.

6.1 UNSTEADY STATE FLOW

In the "steady state" flow, the quantity of fluid entering the system is equal to the quantity leaving the system. In the unsteady state flow, they are usually unequal. New variables, such as time and porosity, are added. The flow of fluid is determined by partial differential equations derived from

1. the material balance equation
2. the continuity equation, and
3. the boundary and initial conditions

Katz [1] gives the most general partial differential equation for unsteady gas flow in radial coordinates as:

$$\frac{\partial^2 p^2}{\partial r^2} + \frac{1}{r}\frac{\partial p^2}{\partial r} = \frac{\mu\phi}{k\bar{p}}\frac{\partial p^2}{\partial t} \tag{6.1}$$

Here, μ, k, and porosity,ϕ, are constants: Variations of Eq. (6.1) defining problems commonly encountered in coal reservoir and production engineering will be solved for given boundary and initial conditions.

6.1.1 Linear Flow of Gas in One Dimension: Gas Produced at a Constant Pressure

Eq. (6.1) becomes simplified as

$$\frac{d^2 p^2}{dx^2} = \frac{\mu\phi}{k\bar{p}}\frac{dp^2}{dt} \tag{6.2}$$

The boundary condition $p = p_w$ is constant at $x = 0$; i.e., the gas well is produced at a constant pressure p_w.

The solution of Eq. (6.2) is

$$\frac{p^2(x,\ t) - p_w^2}{p_e{}^2 - p_w^2} = \text{erfc}\frac{1}{2t_D{}^{1/2}} \tag{6.3}$$

where

$p\ (x,t)$ is the pressure at a distance of x from the well at time t,
p_w is the well-head constant pressure, in psia, and
p_e is the reservoir pressure at $x = \infty$, in psia.

$$\text{Dimensionless time,}\ t_D = \frac{2.634 \times 10^{-4} k\ t\overline{p}}{\mu\phi_c x^2} \tag{6.4}$$

where

k = permeability in md
t = time in hours
$\overline{p}$ = average pressure, $\left(\frac{p_e + p_w}{2}\right)$, in psia
μ = viscosity in cp
ϕ_c = pseudo-porosity for coal
x = distance in feet
erfc is the complementary error function (see Appendix A).

For example: A gas well in a coal seam is producing at a constant pressure of 20 psia. What will be the pressure at a distance of 750 ft after 720 hours of production?

Given: k = 10 md
ϕ_c (pseudo-porosity for coal) = 0.50 (discussed in the next paragraph)
x = distance from the well
Gas viscosity μ = 0.02 cp
z = 1.0
p_e = 500 psia

$$p = \frac{p_e + p_w}{2} = \frac{500 + 20}{2} = 260\ \text{psia}$$

$$t_D = \frac{2.634 \times 10^{-4} \times 10 \times 720 \times 260}{0.02 \times 0.50 \times 750^2} = 0.0876$$

$$\text{Hence:}\ \frac{1}{2 \times (0.0876)^{1/2}} = 1.689$$

$$\text{erfc}\,(1.689) = 1 - \text{erf}(1.689) = 0.0174$$
$$\frac{p^2(x,t) - 20^2}{500^2 - 20^2} = 0.0174$$
$$\text{Hence } p(x,t) = 68.87 \text{ psia}$$

Eq. (6.3) can be used to calculate the spacings, s, of the vertical wells on a longwall panel. The original reservoir pressure needs to be reduced to 100 psi in 5 years at a distance of $s/2$.

6.1.2 Pseudo-Porosity for Coal, ϕ_c

Let us assume that one ton of coal contains 550 ft^3/t of gas at 650 psi at 60°F (a typical value in the Central Appalachian Basin).

One ton of coal has a volume of 25 ft^3. Hence 1 ft^3 of coal contains 22 ft^3 of gas at standard temperature and pressure.

Converting this volume to reservoir conditions:

$$\phi_c = \frac{22 \times 14.7}{650} \times \frac{520}{520} = 0.5 \text{ ft}^3$$
$$\text{Hence } \phi_c = 50\%$$

The coal seam, analyzed as if it were sandstone, has a pseudo-porosity of 50%.

Pseudo-porosity for some US coal seams is as follows:

1. Northern Appalachian Basin: 59%
2. Central Appalachian Basin: 50%
3. Southern Appalachian Basin: 59%
4. San Juan Basin: 55%
5. Illinois Basin: 54%

An average value of s for all coal appears to be 55%.

It is assumed here that the diffusional flow is higher than the Darcy flow in all coal seams, but this is not always the case.

6.1.3 Linear Flow of Gas: Constant Production Rate for a Gas Well

In this case, gas is allowed to flow at a constant rate, Q from the well ($x = 0$).

The solution of Eq. (6.1) is given as per Katz [1]:

$$\frac{p^2(x,t) - p_w^2}{p_w^2} = -\, mp_t \tag{6.5}$$

where

$$m = \frac{8930\ \mu z T Q}{hk\, p_w^2} \tag{6.6}$$

and

$$p_t = \frac{2t_D^{1/2}}{\pi^{1/2} \exp\left(\frac{1}{4} t_D\right)} - \text{erfc}\frac{1}{2t_D^{1/2}} \tag{6.7}$$

The minus sign on the right-hand side in Eq. (6.5) indicates gas production. It becomes positive if gas is injected into the well.

p_t is strictly a function of t_D. Katz [1] provides a table of p_t for values of t_D from 1 to 1000 (see Appendix B).

For values of t_D greater than 1000, Eq (6.7) simplifies to

$$p_t = \frac{1}{2}[\ln t_D + 0.80907] \tag{6.8}$$

Here, p_w, $p(x,t)$,
μ, z, t, and k have the same meaning as in Eq. (6.1).
Q = gas production in MCF/day
h = height of the coal seam in feet
t_D = same as in Eq. (6.4)

6.1.4 Cumulative Gas Flow From a Well Produced at Constant Pressure From an Infinite Reservoir: Radial Coordinates

$$\text{Total cumulative flow, } Q_T = \frac{2\pi\phi}{1000\bar{p}} \times r_w^2 h(\bar{p} - p_w)\left[\frac{520}{T} \times \frac{\bar{p}}{14.7}\right] Q_t \tag{6.9}$$

where $\bar{p}$ = average pressure and
Q_t is a function of t_D.

Values of Q_t can be obtained from tables in the literature. An abridged table is presented in the Handbook of Natural Gas Engineering [1].

Most gas wells in coal seams are produced at constant pressure. In the beginning, a back pressure of 50–100 psi is maintained. After 6 months, the well is produced at atmospheric pressure to maximize gas production.

Eqs. (6.3) and (6.9) give the pressure at a given distance and corresponding cumulative gas production respectively for a well produced at constant pressure. Eq. (6.5) enables the calculation of pressure at a distance if the well is produced at a constant production rate.

6.2 CALCULATION OF EFFECTIVE RESERVOIR PERMEABILITY

Unsteady-state gas flow equations can be used to calculate the effective reservoir permeability using either a draw-down test or a build-up test. In some cases, both tests are done to get an average number for the effective permeability of the coal reservoir.

6.2.1 Draw-Down Test

In this case, a closed well is put on production at a constant rate of flow and the bottom hole pressure (BHP) is measured against time. Since this test is done over a long time, we can combine Eq. (6.5) with (6.8) and write

$$p^2 = -\frac{m}{2}{p_w}^2 \times \ln t + \text{constant} \tag{6.10}$$

As shown in Fig. 3.6, a straight line is obtained if the (BHP)^2 is plotted against the logarithm of t. The slope of the line is $\frac{mp_w^2}{2}$

Substituting for m, from Eq. (6.6) and rearranging, we get

$$K = \frac{1424\mu zTQ}{2h \times \text{slope of the line}} \tag{6.11}$$

where $P_s = \text{BHP}$

Fig. 3.6 shows an example of a draw-down test and calculation of effective reservoir permeability.

6.2.2 Build-Up Test

In this test, a coal gas well that has been producing for a long time is shut in and the build-up of pressure against increments of time is observed.

Since the production rate is altered only once, the pressure–time relationship is expressed as [1]

$$\frac{\bar{p}^2 - p_w^2}{p_w^2} = -\frac{m_1}{2}(\ln t_{D1} - \ln t_{D2}) \tag{6.12}$$

or

$$\bar{p}^2 = m_1\, p_w^2 \left(\frac{\ln(t_f + \Delta t) - \ln \Delta t}{2}\right) + p_w^2 \tag{6.13}$$

where t_f is the length of time the well was flowing at the constant dimensionless rate m_1 prior to shut-in. Since the data is collected only for a short time in this test, Eq. (6.13) can be rewritten as

$$\overline{p}^2 = \frac{m_1 p_w^2}{2} \times \ln\left(\frac{t_f + \Delta t}{\Delta t}\right) + p_w^2 \tag{6.14}$$

A plot of $\overline{p}^2$ against $\ln(t + \Delta t/\Delta t)$ gives a straight line. The slope of the line is equal to $m_1 p_w^2/2$.

Fig. 3.7 shows the plot and calculation of effective reservoir permeability.

Eq. (6.11) can be used again to calculate the effective reservoir permeability.

6.3 STEADY-STATE FLOW OF GAS

When the quantity of fluid entering the well is equal to the quantity exiting from the well, a steady-state condition is achieved.

After the gas production from a well has achieved its peak, production is steady for some time. The drainage radius has not yet reached a finite boundary.

6.3.1 Steady-State Radial Flow in a Vertical Well

When a vertical well of radius r_w is drilled in a coal seam, the steady-state gas flow is given by Eq. (6.15). Smith [2]

$$q = \frac{707.8\ kh(p_e^2 - p_w^2)}{\overline{\mu}\ \overline{z}\ T \ln(r_e/r_w)} \tag{6.15}$$

where

q = cubic ft/day at 60°F and 14.67 psia
k = permeability in darcy
h = thickness in feet
p_e = pressure at external radius, r_e
p_w = pressure at the well radius, r_w
$\overline{\mu}$ = average viscosity
$\overline{z}$ = average compressibility factor
T = temperature in degree Rankine (Fahrenheit + 460)

For liquid flow, Eq. (6.15) becomes

$$Q = \frac{0.03976\ kh\,(p_e - p_w)}{\mu \ln(r_e/r_w)} \tag{6.16}$$

where Q is in CF/day and μ is liquid viscosity.

The rest of units are the same as above.

For example:

Calculate gas and water flow from a well producing steadily under the following conditions:

$k = 0.003$ darcy (3 md)
$h = 40$ ft
$\mu = 0.02$ cp
$z = 0.90$
$T = 60°F$ (+460)
$r_e = 1000$ ft
$r_w = 0.25$ ft
$p_e = 500$ psi
$p_w = 50$ psi

Using Eq. (6.15),

$$Q = \frac{707.8 \times (0.003)40(500^2 - 50^2)}{0.9 \times 520 \times (0.02) \times \ln\left(\frac{1000}{0.25}\right)}$$

$$= 270.9 \text{ MCFD}$$

The above conditions describe a typical well drilled into a thick seam with good permeability. The well is produced at a constant pressure of 50 psi. Similarly, using Eq. (6.16), the water flow can be calculated as 46.2 bbl/day.

6.3.2 Solving Practical Problems in Reservoir Engineering

Eq. (6.15) can be used to solve many practical problems in reservoir engineering. Only the most common cases will be discussed here.

6.3.2.1 Case 1. Impact of Hydrofracking a Vertical Well

The process will be discussed in Chapter 8, Hydraulic Fracking of Coal, but basically, the well radius is extended bilaterally to about 500 ft. The increase in gas production can be calculated as follows.

Let us assume
original well radius = r_{w1}
fracked well radius = r_{w2}
Eq. (6.15) can be used to write

$$\frac{Q_2}{Q_1} = \frac{\ln r_e/r_{w1}}{\ln r_e/r_{w2}} \tag{6.17}$$

Assuming $r_{w1} = 0.25$ ft, Q_1 is initial production
$r_{w2} = 500$ ft, and
$r_e = 1000$ ft Q_2 is the production after hydrofracking

$$\frac{Q_2}{Q_1} = 12$$

That is, gas production can be increased by a factor of 12.

Illustrative problem:

Calculate the initial gas flow and the gas flow after successful hydrofracking, given

1. Original well
   $k = 0.01$ darcy
   $h = 6$ ft
   $\mu = 0.02$ cp
   $z = 1.0$
   $T = 520°R$
   $r_e = 1000$ ft
   $r_{w1} = 0.25$ ft
   $p_e = 215$ psi
   $p_w = 15$ psi
   $Q_1 = 22{,}657\ \text{ft}^3/\text{day} = 22.65$ MCFD
2. If this well is fracked to increase the radius to 500 ft

$$Q_2 = 12Q_1 = 272\ \text{MCFD}$$

The above conditions depict a typical well in the Central Appalachian Basin.

6.3.2.2 Production From a Vertical Well Hydrofracked in Several Coal Seams

A typical vertical well in the Central Appalachian Basin is fracked in multiple coal seams. All coal seams can be hydrofracked in a single operation to enhance gas production.

The combined production would be the sum of production Q_i from a coal seam of thickness h_i and permeability k_i.

$$\text{Total production} = \sum_{i=1}^{n} Q_i = \frac{707.8(p_e^2 - p_w^2)}{\mu z T \ln(r_e/r_w)} \sum_{i=1}^{n} k_i h_i \tag{6.18}$$

It is assumed that each coal seam is hydrofracked identically.

Illustrative example:

Assume that three coal seams in the Central Appalachian Basin are identically fracked with gelled water to create a drainage radius of 500 ft: K_1, K_2, K_3 are 10, 10, and 10 md. The seam thicknesses are 5, 4, and 3 ft, respectively. The reservoir pressure $p_e = 500$ psia and the temperature is 60°F.

$$\text{Total production} = \frac{707.8\,(500^2 - 15^2)}{(520)\,(0.02)\ln(500/0.25)} \left(\frac{10 \times 5 + 10 \times 4 + 10 \times 3}{1000} \right)$$
$$= 268.6\ \text{MCFD}$$

Eq. (6.18) can be used to calculate gas production from gas wells in the Central and Southern Appalachian Basins where a single well is completed in multiple coal seams.

6.4 PRODUCTION DECLINE

When the drainage radius of a gas well reaches a boundary (end of the reservoir) or it interferes with another nearby well, gas production begins to decline. For oil and gas fields, Arps [3] first identified the three types of production declines: exponential, harmonic, and hyperbolic curves (Fig. 6.2). The hyperbolic decline curve can be considered a generalized model because the other two curves can be derived from it.

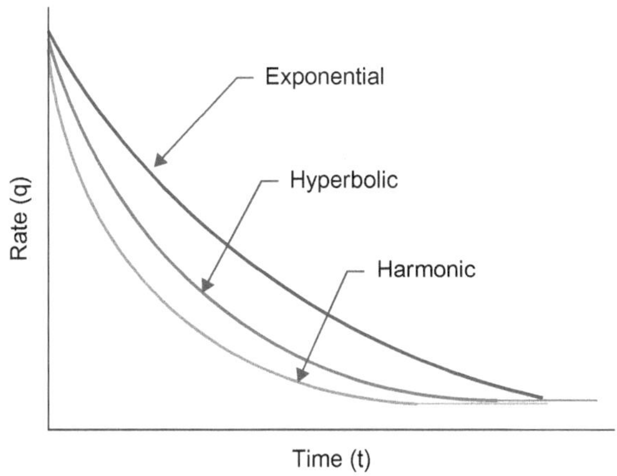

Figure 6.2 Decline curves for gas production from sandstone.

6.4.1 Exponential Decline

The exponential decline can be mathematically expressed as

$$q_t = q_i e^{-dt} \tag{6.19}$$

where

q_i is the initial production rate
q_t is the production at time t
d is the rate of decline
t is the time

Taking the logarithm of both sides,

$$\ln q_t = \ln q_i - d_i t \tag{6.20}$$

Hence, a plot of $\ln q_t$ against time t will yield a straight line with a slope of d_i and an intercept of $\ln q_i$.

Exponential decline is the most commonly used decline curve for natural gas production wells.

6.4.2 Harmonic Decline

This model is not commonly used, but is useful when a plot of cumulative production, $Q_{p,}$ against $\ln t$ is linear.

Mathematically, we can modify Eq. (6.20) and rewrite

$$\ln q_t = \ln q_i - d_i \frac{Q_p}{q_i} \tag{6.21}$$

A plot of $\ln q_t$ against Q_p will yield a straight line. The gradient will be d_i/q_i and the intercept will be $\ln q_i$. Q_p is the total, cumulative gas production.

6.4.3 Hyperbolic Decline

This is the generic form of all production decline curves.

Mathematically,

$$q_t = \frac{q_i}{(1+nd_i t)^{1/n}} \tag{6.22}$$

where $0 < n < 1$.

The type curve overlay method can be used to determine q_i, d_i, and n. It is also proposed that a log–log type curve to match the decline curve can be used to determine the q_i, d_i, and n.

The best way to determine these parameters is to use weighted residual nonlinear regression.

Besides the above types of decline curves, three additional curves have recently been proposed that fit production decline well. They will be discussed only briefly.

6.4.4 Power Law Exponential Decline Curve

Ilk [4] proposed this by modifying Arp's exponential decline curves. It was developed especially for gas production decline from tight sands.

Mathematically, the law can be written as

$$q = q_{\mathrm{i}}\, \mathrm{e}^{\left[D\infty t - \frac{D_{\mathrm{i}}}{n} t^{n}\right]} \tag{6.23}$$

It resembles Arp's exponential decline, but yields better fit to gas production from tight reservoirs. $D\infty$ denotes the final value of the decline rate.

6.4.5 Stretched Exponential Decline

Valko and Lee [5] proposed a slightly different exponential decline.

Mathematically,

$$q_{\mathrm{t}} = q_{\mathrm{i}} \exp\left(1 - \left(\frac{t}{\tau}\right)^{n}\right) \tag{6.24}$$

where q_{i} is initial production rate, and τ is a characteristic time that corresponds to 63.3% of total production.

6.4.6 Power Law Decline

Thakur [6,7] presented several plots of actual gas production from coal that was drilled vertically and horizontally.

Mathematically,

$$Q_{\mathrm{t}} = At^{n} \tag{6.25}$$

where

Q_{t} is the cumulative production at time t

A is the initial production rate

t is time in days or months

n is a characteristic constant that depends on well geometry and coal properties

Taking the log of both sides in Eq. (6.25),

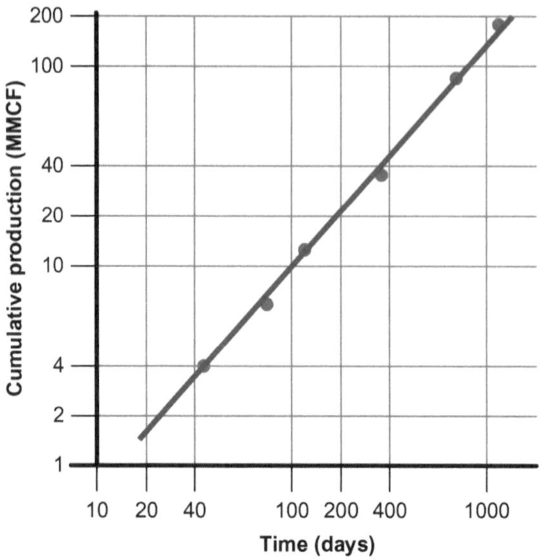

Figure 6.3 Cumulative gas production from a typical horizontal borehole.

$$\ln Q_t = \ln A + n \ln t \tag{6.26}$$

Plotting ln Q_t against ln t on log–log paper yields a straight line. The gradient is the decline exponent and is usually less than 1.00. The intercept gives the logarithm of the initial production rate.

Eq. (6.26) was plotted for various cases of gas production from coal, shown in Figs. 6.3–6.5.

6.4.6.1 *Case 1*

Fig. 6.3 shows cumulative gas production from a 1000 ft horizontal well drilled in the Pittsburgh seam of the Appalachian Basin. It produced 36 MMCF in the first 300 days with an average specific production of 12 MCFD/100 ft of the borehole. The production decline exponent, n, is 0.8. Additional data on gas production from horizontal well is provided by Thakur [7].

6.4.6.2 *Case 2*

Fig. 6.4 shows cumulative gas production from a vertical well hydrofracked in several coal seams. Cumulative production in 6 years was 662 MMCF. The logarithmic plot again yields a straight line. The production decline exponent, n, is 0.81.

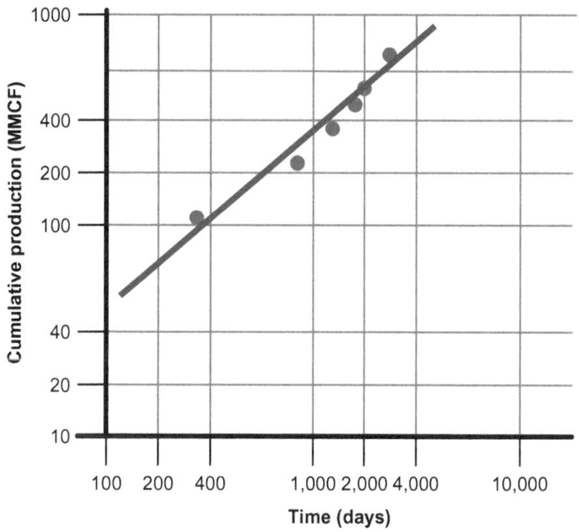

Figure 6.4 Cumulative gas production from a typical vertical well in multiple coal seams.

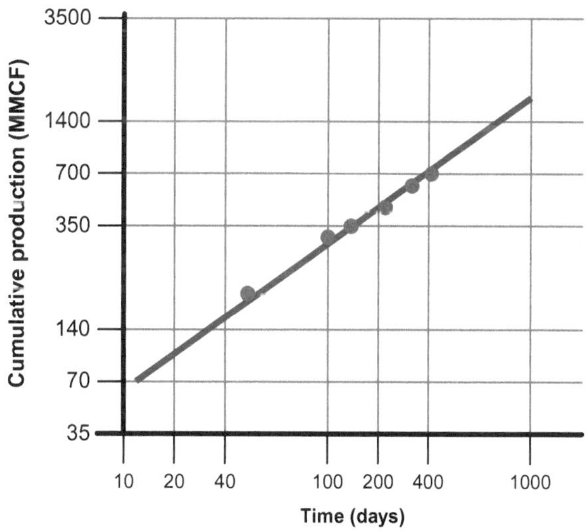

Figure 6.5 Cumulative gas production from gob areas of mines.

6.4.6.3 Case 3

Substantial gas production is realized from coal mines where longwall mining has been done. A mined-out area is called a "gob." There are usually many coal seams overlying and underlying the mined-out coal seams. As they get destressed due to subsidence and heaving of the floor respectively, they release gas contained in them into the gob.

Fig. 6.5 shows a typical gas production decline graph for the Pocahontas No. 3 seam in the Central Appalachian Basin. This log–log plot of cumulative gas production against time again yields a straight line. The production decline exponent, n, in this case is approximately 0.7, showing a slower rate of decline. This is mainly due to the coal being broken and not a solid mass as in the previous two cases. Greater details on this are provided by Thakur [6].

REFERENCES

[1] Katz DL. Handbook of natural gas engineering. New York, NY: McGraw-Hill Book Company; 1958. p. 403–20.

[2] Smith RV. Practical natural gas engineering. Tulsa, OK: Pennwell Publishing Company; 1990. p. 97–107.

[3] Arps JJ. Analysis of decline curves. SPE; 1945. p. 160–228.

[4] Ilk D, et al. Exponential vs hyperbolic decline in tight sands – understanding the origin of implications for reserve estimate using Arp's decline curves, SPE Paper 116731, Denver, Colorado, 2008.

[5] Valko PP, et al. A better way to forecast production from unconventional gas wells, SPE Paper 134231. 2010.

[6] Thakur PC. Methane control in longwall gobs, Longwall and Shortwall Mining, State-of-the-Art, AIME, 1980, 81–6.

[7] Thakur PC. Methane flow in the Pittsburgh coal seam, USA, the 3rd International Mine Ventilation Congress. Howes MS, Jones MJ, editors. Harrogate, England, 1984, p. 177–82.

CHAPTER 7

Fluid Flow in Pipes and Boreholes

Contents

Abstract

The third and final phase of fluid flow from coal seams is the flow in pipes and cased boreholes. Unlike the diffusional flow from coal particles and darcy flow in coal matrices, the size of the conduits can be designed so that they do not restrict the flow. The four most pertinent cases of fluid flow are discussed: (1) flow of water (or other liquids) in pipes, (2) flow of gases in pipes, (3) flow of sand/coal slurry in pipes, and (4) flow of gases in vertical wells. In each case, pressure losses are calculated to determine the horsepower needed for the job or alternatively to design the size of the conduit to minimize the horsepower required. Dimensional analysis of multiple variables is done to express friction as a function of the Reynolds number and the roughness of the pipe for laminar flow. For turbulent flow only the roughness of the pipe is significant. Various equations are provided to calculate the friction factors. The minimum velocity to transport sand or coal cuttings in pipes or an annulus is calculated for various pipe sizes. The friction coefficient for slurries can be derived if the velocity, volumetric concentration, and drag coefficients of particles are known. Horsepower to drive a water pump that can enable drilling 3000 ft long boreholes is calculated. Finally, compressor horsepower for surface transport of gases is calculated. Effective diameters of noncircular pipes are also defined.

Advanced Reservoir and Production Engineering for Coal Bed Methane.
DOI: http://dx.doi.org/10.1016/B978-0-12-803095-0.00007-7

The third and final phase of fluid flow from coal seams is the flow in pipes and boreholes. Unlike the diffusional flow from coal particles and darcy flow in coal matrices, the size of the conduits can be designed so that they do not restrict the gas flow.

Fluid flow in pipes is a vast subject. Only those flow regimes that are most pertinent for gas production engineering will be considered. All flows are fully turbulent and the friction coefficient is regarded as a constant. Once the gas reaches the surface, it is compressed and conducted through pipes to a processing plant. After processing, the gas is again compressed to transport it to the market through pipelines. Hydraulic fracturing of coal or horizontal drilling of coal uses, respectively, water and sand slurry and coal slurry transportation. It is, therefore, essential to know the design principles for these operations.

7.1 DERIVATION OF THE BASIC EQUATION

Basic equations for fluid flow were initially developed for ideal (frictionless) fluids that may be compressible or incompressible [1]. In the realm of coal bed methane production, all fluids are real fluids, i.e., they have viscosity and they create friction. Moreover, all the flow regimes are mainly turbulent. The flow of water, gases, and solids can create a large number of situations, but this chapter will deal with only four types of flows in pipes:

1. Flow of water in pipes.
2. Flow of gases in pipes.
3. Flow of water and sand or coal in a slurry form.
4. Flow of gases in vertical wells.

7.2 FLOW OF WATER IN PIPES

The solutions of practical water flow problems are derived from energy principles, the equation of continuity and the equation of fluid resistance. Resistance to flow in pipes is offered not only by frictional losses in long pipes but also by pipe fittings, such as bends and valves, which create turbulence and hence dissipate energy.

Early experiments (by Darcy, c.1850) on the flow of water in pipes indicated that the pressure loss was directly proportional to the length of the pipe and the velocity head $\left(\frac{v^2}{2g}\right)$ and inversely proportional to the diameter, d. Mathematically, it can be expressed as

$$h = \frac{\lambda l v^2}{2gd} \tag{7.1}$$

where

h is the head loss over a distance, l in ft of the fluid
l is the length of the pipe in ft
d is the pipe diameter in ft
g is the acceleration due to gravity (32 ft/sec^2) and
λ is a coefficient of proportionality, commonly called the friction factor.
It is dimensionless. v is the velocity, ft/sec.

7.2.1 Determination of λ in Eq. (7.1)

In order to calculate the head (pressure) loss in Eq. (7.1), the only thing not known is the friction factor, λ.

Dimensional analysis will be used to determine the variables that can be used to predict λ. When a viscous fluid flows in a pipe, the frictional stress τ_o is dependent on the following variables only:

v is the velocity of the fluid in ft/second (L/T)
d is the pipe diameter in ft (L)
ρ is the density of the fluid in lb/ft^3 (M/L^3)
μ is the viscosity of the fluid in poise (M/LT) and
e is the pipe roughness in ft (L).

Mathematically

$$\tau_o = F(v, d, \rho, \mu, e) \tag{7.2}$$

Dimensional analysis converts Eq. (7.2) into

$$\frac{M}{T^2 L} = \left(\frac{L}{T}\right)^a (L)^b \left(\frac{M}{L^3}\right)^c \left(\frac{M}{LT}\right)^d (L)^e$$

Comparing the power of mass (M), length (L), and time on both sides, for

M: $1 = c + d$
L: $-1 = a + b - 3c - d + e$
T: $-2 = -a - d$ or $2 = a + d$

We can eliminate three unknowns by converting a, b, and c into d and e.

$$a = 2 - \vec{d}, \quad b = -(d + e), \quad \text{and} \quad c = 1 - d.$$

Thus Eq. (7.2) can be rewritten as

$$\tau_o = c\, v^{2-d}\, d^{-de}\, \rho^{1-d}\, \mu^d\, e^e \tag{7.3}$$

or

$$\tau_o = c\left(\frac{\mu}{v d \rho}\right)^d \left(\frac{e}{d}\right)^e \rho \cdot v^2$$

where c is constant of proportionality

Eq. (7.3) shows that frictional losses in a pipe are basically a function of two variables:

$\frac{v d \rho}{\mu}$ is defined as the Reynolds number, R, and $\left(\frac{e}{d}\right)$ is the roughness factor.

Stanton [2] and Nikuradse [3] have carried out extensive research on measuring λ for Reynolds numbers ranging from 10^3 to 10^6 and $\left(\frac{e}{d}\right)$ ranging from 1×10^{-4} to 1×10^{-2}. See these works for details.

The roughness, e, for various commercial pipes is shown in Table 7.1.

Colebrook [4] studied the roughness of many pipes and came up with a single equation that can be used very conveniently.

$$\frac{1}{\lambda} = 2 \log \frac{d}{e} + 1.14 - 2 \log \left[1 + \frac{9.28}{R\left(\frac{e}{d}\right)\sqrt{\lambda}}\right] \tag{7.4}$$

The value of λ is obtained by several iterations. Eq. (7.4) has been made user-friendly by the Moody [5] diagram, which shows the value of λ against

Table 7.1 Roughness for various pipes

Type of pipe	*e* (inches)
Wrought iron	0.0017
Well tubing/line pipe	0.0007
Cast iron	0.0050
Galvanized iron	0.0060
Uncoated cast iron	0.0100
Wood	0.007–0.036
Concrete	0.012–0.12
Riveted steel	0.035–0.35

varying Reynolds numbers and different $\left(\frac{e}{d}\right)$ ratios ranging from 10^{-5} to 10^{-1}. Mostly, the λ values range from 0.01 to 0.09, representing very smooth to wholly rough pipes. For fully turbulent flow in smooth pipes, Vennard [1] provides another equation for λ that may be easier to use.

$$\frac{1}{\lambda} = -0.80 + 2.0 \log \mathrm{R}\sqrt{\lambda} \tag{7.5}$$

When the flow becomes completely turbulent—that is, beyond the transition zone—the frictional coefficient is no longer a function of Reynolds number, but becomes a function of e/d only. The friction factor in this region of flow is completely independent of the physical properties of the flowing fluid. For fully turbulent flow λ is expressed by an equation obtained experimentally by Nikuradse [6]:

$$\frac{1}{\lambda} = 2 \log \frac{d}{e} + 1.14 \tag{7.6}$$

Thus for a 6-inch diameter cast iron pipe with roughness of 0.005 inches

$$\frac{d}{e} = \frac{6}{0.005} = 1200$$

$$\text{Hence, } \frac{1}{\sqrt{\lambda}} = 2 \log 1200 + 1.14 = 7.3$$

$$\text{or } \lambda = \left(\frac{1}{7.3}\right)^2 = 0.0188$$

An example:

In drilling a horizontal borehole, the drill motor needs 75 gpm water. Calculate the head (pressure) loss for 3 000 ft of 3-inch diameter drill pipe. The roughness of the pipe, e, is 0.006.

Step 1. Calculate the fluid velocity.

Q, fluid flow rate = 75 gpm = 10 ft^3/min = 0.167 ft^3/second

A, cross-section of pipe $= \frac{\pi}{4}\left(\frac{3}{12}\right)^2 = 0.049\ \text{ft}^2$

Hence velocity, $V = \frac{Q}{A} = \frac{0.167}{0.049} = 3.41$ ft/second

Step 2. Calculate the Reynolds number, R.

$$\mathrm{R} = \frac{Vd}{(\mu/\rho)} = \frac{3.41 \times 0.25}{1.217 \times 10^{-5}} = 70,000$$

(μ/ρ is the kinematic viscosity of water $= 1.217 \times 10^{-5}\ \text{ft}^2$/second)

Hence the flow is fully turbulent.

Step 3. Calculate λ from Eq. (7.6).

$$\frac{1}{\sqrt{\lambda}} = 2\log\left(\frac{3}{0.006}\right) + 1.14$$

This gives $\lambda = 0.0234$.

Step 4. Calculate head loss using Eq. (7.1).

$$h = \frac{0.0234 \times 3000\,(3.41)^2}{2 \times 32 \times 0.25}\ \text{feet of water}$$

$$= 51\ \text{ft} = 22.2\ \text{psi}$$

7.3 GAS FLOW IN HORIZONTAL PIPELINES

Frictional losses in transporting large volumes of gases in pipelines must be accurately determined to design the compressors. The following assumptions are made:

1. The kinetic energy change is negligible.
2. The flow is steady and isothermal.
3. The flow is horizontal.
4. There is no work done by the flowing gas.

The equation governing the pressure loss for this case is

$$\int_1^2 v\,dp + \int_1^2 \frac{\lambda l v^2}{2gd}\,dl = 0 \tag{7.7}$$

One of the earliest equations that related the volume of gas flow Q to pressure losses was by Weymouth [7] and is given below:

$$Q = 3.22\frac{T_o}{P_o}\left[\frac{(P_1^2 - P_2^2)d^5}{GTL\lambda Z}\right]^{0.5} \tag{7.8}$$

where

Q = gas flow measured at T_o & P_o, in cubic ft/hr
L = length of pipeline, in miles
d = internal diameter, in inches
p = pressure, in psia
G = gas specific gravity (air = 1)
T = average line temperature, °R
Z = average compressibility factor
λ = friction factor, dimensionless

Weymouth assumed that λ varied as a function of pipe diameter and can be estimated by Eq. (7.9):

$$\lambda = \frac{0.032}{d^{1/3}} \tag{7.9}$$

Eq. (7.8) then changes into Eq. (7.10) when we substitute for λ:

$$Q = 18.062 \frac{T_o}{P_o}\left[\frac{(P_1^2 - P_2^2)\, d^{16/3}}{GTLZ}\right]^{0.5} \tag{7.10}$$

There are many other equations that relate gas flow rate with pressure loss. Reference can be made to Katz [8] for additional details.

An example:

Find the diameter of a pipeline to deliver 100 MMCFD to a distance of 40 miles given the following conditions:

T_0 = base temperature 60°F = 520°R

P_0 = 15 psia

P_1 = 1000 psia

P_2 = 300 psia

G = 0.6

T = average temperature of the line, 510°R

L = 40 miles

Z = 1.00

First calculate

$$Q = \frac{100 \times 10^6}{24} = 4.17 \times 10^6 \text{ ft}^3/\text{hour}$$

Hence, $$4.17 \times 10^6 = 18.062 \frac{520}{15}\left[\frac{(1000^2 - 300^2)d^{16/3}}{0.6 \times 510 \times 40}\right]^{0.5}$$

or $$d^{16/3} = 596,566$$

Hence, $$d = 12.1 \text{ inch diameter.}$$

To be on the safe side, one would go with 16-inch diameter pipe. Most commonly, the pipeline diameter is underestimated.

7.4 HYDRAULIC TRANSPORT OF SOLIDS IN WATER

In the process of drilling for gas production from coal seams, the drill cuttings need to be transported in suspension in water through an annulus. To design an economical pumping system one needs to know the flow rate, concentrations of solids, pipe size, minimum transport velocity, pump size, and horsepower needed.

7.4.1 The Concentration of Solids

The drill motor drills at a maximum rate of 10 ft/min, and the borehole diameter is 3.5 in. Hence the drill cuttings are created at a rate of 0.668 ft^3/min (53.4 lbs. of coal).

The pumping rate should be high enough to carry the coal in suspension and such that the concentration of solids by weight does not exceed 50%. The drill motor (run by high-pressure water) needs 75 gpm (10 ft^3/min) for proper operation.

The volumetric concentration of solids is $\frac{0.668}{10} \times 100 = 6.68\%$. The weight concentration of solids is $\frac{53.4}{624} = 8.55\%$. These numbers are well below the limiting volumetric and weight concentrations. One foot of the slurry equates to $(0.434 \times 1.0855) = 0.471$ psi.

The slurry density is 67.7 lbs/ft^3.

7.4.2 Minimum Transport Velocity

The velocity of the water in the pipe should be optimum. If the velocity is too low, the solids will settle out, resulting in blockage of the pipe and the water flow. If the velocity is too high, friction losses will be high. Power consumption will then be excessive and wear and tear on the pumps and pipeline will be more severe. The problem then is to determine the minimum safe velocity.

Most drill cuttings are less than 3/8 inches in size with a mean size of 1/8 inch (approximately 3.2 mm). Durand [9] provides an approximation of the minimum transport velocity as

$$V_L = F_L \sqrt{2gd \frac{S - S_L}{S_L}} \tag{7.11}$$

where

V_L is the minimum (limiting settling) velocity in ft/second

F_L is a factor dependent on particle diameter and concentration for sand/coal. For cuttings in this case a value of 1.34 is adequate.

g is the acceleration due to gravity

d is the pipe size, in ft

S is the specific gravity of the solids

Table 7.2 Minimum transport velocity

Pipe diameter (inches)	**Minimum transport velocities (ft/second)**	
	Sand	**Coal**
3	6.89	2.94
4	7.95	3.39
6	9.74	4.15

dp (mean particle diameter): 1/8″ (3 mm).
Sandstone specific gravity = 2.65.
Coal specific gravity = 1.300.

S_L is the specific gravity of water, typically 1.00

While drilling in a coal seam one is likely to run into shale or sandstone bands, hence the minimum velocity is always designed to carry the heaviest particles (in this case sandstone) in suspension. Table 7.2 shows the minimum transport velocity for sandstone and coal particles in various pipe sizes.

The area of the annulus between the borehole (3.5 inch diameter) and the outside diameter of the drill pipe (2.75 inches) is equal to 0.0256 ft^2. At 75 gpm, it results in a fluid velocity of 6.5 ft/second. This is enough to carry coal particles but barely enough to carry sandstone particles in suspension.

7.4.3 Determination of the Frictional Coefficient, λ_S, for Slurries

The first step is to calculate the λ_w (frictional coefficient for water) as discussed earlier and then use Durand's [9] empirical formula to calculate λ_s, the frictional coefficient for the slurry.

$$\lambda_S = \lambda_W \left[1 + 82 \left(\frac{gD}{V^2} \; \frac{\rho - \rho_W}{\rho_W} \right)^{3/2} \frac{c_V}{c_D^{3/4}} \right] \tag{7.12}$$

Using the values of v, c_v, and c_D (the value is equal to 0.44 for most solid particles) as worked out earlier,

$$\lambda_S = \lambda_W(1.003) = 0.0234 \times 1.003$$
$$= 0.0234$$

The pressure loss in the pipeline can now be calculated using Eq. (7.1), but the head loss is now in terms of feet of slurry.

7.4.4 Calculation of Pressure Loss

Assuming that the depth of the horizontal borehole is 3000 ft, friction loss equals

$$h_s = \frac{\lambda / V^2}{2gd}$$

$$= \frac{0.0234 \times 3000 \times (6.5)^2}{2 \times 32 \times (d_2 - d_1)}$$

where

d_2 is the inside diameter of the borehole and
d_1 is the external diameter of the drill pipe.
Hence,

$$d_2 = 3.5 \text{ in.} : d_1 = 2.75 \text{ in.}$$

$$h_s = \frac{0.0234 \times 3000 \times (6.5)^2}{2 \times 32 \left(\frac{3.5 - 2.75}{12} \right)} = 741 \text{ ft.}$$

One foot of the slurry is equal to 0.47 psi. Hence, the frictional loss is 349.2 psi.

Total head loss in the pumping system

$$\begin{aligned} &= \text{pressure loss in drill pipe} \\ &+ \text{pressure loss in drill (water} - \text{driven) motor (assume 200 psi)} \\ &+ \text{pressure loss on the annulus} \\ &= 22.2 + 200 + 349.2 = 571.4 \text{ psi} \end{aligned}$$

To cover all other losses, such as valves, bends, and fittings, the total designed pressure loss was put at 900 psi. This also gives a bit of reserve if the drill motor needs a higher pressure to drill through some material harder than coal.

7.4.5 Calculation of Pump Horsepower

Having decided the fluid flow rate at 75 gpm and total pressure loss of 900 psi, one can now calculate the horsepower for the pump by Eq. (7.13).

$$\text{Horsepower} = \frac{(\text{Pressure, psi}) \times (\text{Flow, gpm})}{1714\,\eta} \tag{7.13}$$

where

η is the pump and drive train efficiency, assumed to be 0.8.

In this case horsepower needed $= \dfrac{900 \times 75}{1714 \times 0.8} = 49.22$

A 50 hp motor would be adequate. This is an actual case for a horizontal drilling system designed to drill in coal to a depth of 3000 ft. The problem is often made more difficult by the loss of fluid in the coal formations. An optimum size of the pump and motor is determined by experience.

7.5 FLOW OF GAS IN A VERTICAL WELL

To allow for vertical height, Eq. (7.7) is modified as follows:

$$\int_1^2 v\,dp + \int_1^2 \frac{\lambda\, v^2 dl}{2gd} + \frac{g\,\Delta x}{g_L} \tag{7.14}$$

The main assumptions are the same, except that the flow is vertical. Smith [10] has derived a solution for Eq. (7.14) as

$$Q = 200,000\left[\frac{d^5}{GTZ\lambda} \times (P_2^2 - e^S\,P_1^2)\,\frac{S}{S-1}\right]^{0.5} \tag{7.15}$$

Substituting for $\lambda = \frac{0.032}{d^{1/3}}$ from Eq. (7.9) we get

$$Q = 1.118 \times 10^6\left[\frac{d^{16/3}}{GTZ} \times (P_2^2 - e^S\,P_1^2)\,\frac{S}{e^S - 1}\right]^{0.5} \tag{7.16}$$

where

Q = volume flow rate at 14.65 psia and 60°F, in ft^3/day

Z = average compressibility

T = Average temp, in °R

d = well diameter in inches

P_2 = bottom hole pressure

P_1 = well head pressure

$e = 2.7183$

$S = \frac{0.0375\ Gx}{T\ Z}$

G = gas specific gravity

x = difference in elevation, in ft

An example:

Calculate the gas flow from a 2000 ft deep well with 5 inch diameter casing with the following conditions:

$G = 0.6$

$T = 520°R$

$Z = 1.00$

$P_2 = 700$ psia

$P_1 = 15$ psia

$$S = \frac{0.0375 \times 0.6 \times 2000}{520 \times 1} = \frac{45}{520} = 0.0865$$

$$\text{Hence,}\quad Q = 1.118 \times 10^6 \left[\frac{(0.416)^{16/3}}{0.6 \times 520 \times 1} \left(700^2 - 1.09 \times 15^2\right) \frac{0.0865}{1.09 - 1} \right]^{0.5}$$

$$= 4.19 \text{ MMCFD}$$

7.6 CALCULATION OF HORSEPOWER FOR GAS COMPRESSION

Since all gas transportation on the surface is done by compressing the gas, it is necessary to know the horsepower needed for a specific flow rate and pressure. Traditionally, enthalpy-entropy diagrams are used for rigorous calculations of horsepower needed for compression. Fortunately, a simple equation was developed by Joffe [11] that agrees very well with enthalpy-entropy diagrams as shown in Eq. (7.17):

$$-W = \frac{K}{K-1} \frac{53.241 T_1}{G} \left[\left(\frac{P_2}{P_1} \right)^{\frac{Z(K-1)}{K}} - 1 \right] \tag{7.17}$$

The above equation can be rewritten in field units to compress 1 MMCF/day at 60°F and 14.65 psi as

$$-W = 0.08531 \frac{K}{K-1} T_1 \left[\left(\frac{P_2}{P_1} \right)^{\frac{Z(K-1)}{K}} - 1 \right] \tag{7.18}$$

where

W = work required to compress real gas, in ft-lb/lb

T_1 = temp at inlet in °R

$K = C_p/C_v$ for gas at inlet

Z = compressibility factor

P_1 = suction pressure, in psia

P_2 = discharge pressure, in psia

An example:

Calculate the adiabatic horsepower required to compress 1 MMCF/day of 0.6 specific gravity coal seam gas at 100 psi and 80°F to 500 psia. Use Eq. (7.18) to calculate the horsepower

where

$K = 1.28$

$T_1 = 540°R$

$\frac{P_2}{P_1} = 5$

Z (average) = 0.985

Hence,

$$W = 0.08531\left(\frac{1.28}{0.28}\right)1.28\ 540\left(5^{0.985\left(\frac{0.28}{1.28}\right)} - 1\right)$$

$$= 87 \text{ hp.}$$

Allowing for an efficiency of 0.8, the motor size should be 109 hp.

Typically, the gas pressure is increased to 400–500 psi in one stage. Additional stages would be needed for higher pressures. Horsepower for the second stage compression can be similarly obtained. The total horsepower needed would be the sum of the horsepower needed in each stage.

7.7 EFFECTIVE DIAMETER OF NONCIRCULAR PIPES

Most often the gas flow in a typical gas well is through an annular space—the space between the inside diameter of the well and the outside diameter of the tubing used to get the water out. Since it is not practical to install a down-hole pressure gauge in the annulus, the pressure losses are always obtained by calculation.

All equations for pressure loss mentioned earlier can still be used if the diameter is replaced by an effective diameter, $d_{\text{eff}} = (d_2 - d_1)$,

where

d_1 is the external diameter of the tubing and

d_2 is the inside diameter of the gas well.

The friction factor can be obtained by modifying Eq. (7.6) as

$$\frac{1}{\sqrt{\lambda}} = 2\log\frac{(d_2 - d_1)}{e} + 1.14 \tag{7.19}$$

For noncircular pipes, the Reynolds number, R, is based on hydraulic diameter, D_h, defined as

$$D_h = \frac{4\,A_c}{P} \tag{7.20}$$

where A_c is the cross-sectional area of the noncircular pipe and P is the wetted perimeter.

Thus for a circular pipe,

$$D_h = \frac{4\left(\frac{\pi\,D^2}{4}\right)}{\pi\,D} = D$$

For a square duct with sides equal to a,

$$D_h = \frac{4\,a^2}{4\,a} = a$$

For a rectangular duct with sides equal to a and b,

$$D_h = \frac{4\,ab}{2(a+b)} = \frac{2\,ab}{a+b}$$

It is generally agreed that
for Laminar flow $R \leq 2300$,
for transitional flow $2300 < R < 4000$, and
for turbulent flow $R \geq 4000$.

REFERENCES

[1] Vennard J. Elementary fluid mechanics. John Wiley and Sons, Inc.; 1961. p. 570.
[2] Stanton TE. Similarity of motion in relation to the surface friction of fluids. Trans R Soc Lond A 1914;214.
[3] Nikuradse J. Stromungs gesetze in Rauhen Rohren. VDI-Forschungsheft 1933;361. [in German].
[4] Colebrook CF, White CM. The reduction of carrying capacity of pipes with age. J Inst Civil Eng Lond 1937;7:99.
[5] Moody LF. Friction factors for pipe flow. Trans ASME 1944;66:671.
[6] Nikuradse J. VDI-Forschungsheft, No. 356, 1938 [in German].
[7] Weymouth TR. Problems in natural gas engineering. Trans ASME 1912;34:185.
[8] Katz D, et al. Handbook of natural gas engineering. McGraw-Hill Book Company; 1958. p. 304–9.
[9] Durand R. Hydraulic transportation of coal and other solid materials in pipes. London: Colloquim of National Coal Board; 1952.
[10] Smith RV. Determining friction factors for measuring productivity of gas wells. Trans AIME 1950;189:73.
[11] Joffe J. Gas compressors. Chem Eng Prog 1951;47:80.

CHAPTER 8

Hydraulic Fracking of Coal

Contents

Abstract

Hydraulic fracking is by far the most-used technique for coalbed methane production today. Under ideal circumstances it yields good gas production. The process involves drilling into the coal seam and casing the gas well; minifrac of the coal to determine its hydrofracking parameters; and finally pumping a calculated quantity of a fluid (water, gelled water, or cross-linked gel) mixed with sand to create a vertical, 500–1000 ft long, bilateral fracture that produces gas. Minifrac is a small-scale version of hydrofracking that yields important information, such as instantaneous shut-in pressure, frac gradient, reservoir (pore) pressure, permeability, and fracture extension pressure. Theoretical calculations of the length, width, and height of fracture as well as its direction are done to estimate the fracture volume and efficiency. Three different models of fracture growth are discussed: (1) the Perkins and Kern model for water fracture of coal, (2) the Geertsma and deKlerck model for a cross-linked gel (highly viscous) fluid fracture, and (3) a radial model for horizontal fractures. Fracture designs

Advanced Reservoir and Production Engineering for Coal Bed Methane.
DOI: http://dx.doi.org/10.1016/B978-0-12-803095-0.00008-9

for three cases are presented: (1) water fracture of coal at medium depth, (2) N_2-foam fracture of coal at medium depth, and (3) slick water fracture of deep and thick coal seams that require a high rate of fluid flow (approaching 100 bpm). Finally, fracture pressure analysis is presented. It generally has four modes (only three in coal seams). Mode I with a slope of ⅛ to ¼ indicates unrestricted linear growth with restricted height. Mode II is flat and indicates fracture extension with moderate height growth. Mode III with a steep slope of 45–63 degrees indicates restricted extension, mainly due to the dissipation of fluid (leak-off) or net pressure that creates the fracture.

At the current state of the technology, there are basically four techniques for gas production from coal seams:

1. Hydraulic fracking of vertical wells in coal.
2. Horizontal drilling in shallow coal seams without fracking.
3. Hydraulic fracking of horizontal boreholes drilled in deep coal seams.
4. Vertical wells in thick, shallow coal seams without hydrofracking.

The author developed the horizontal drilling technology over the period 1974–84, and then perfected hydraulic fracturing of coal over the period 1984–94. The wealth of information about this will be condensed in Chapter 9, Horizontal Drilling in Coal Seams, and this chapter, respectively.

8.1 THE PROCESS OF HYDROFRACKING

In coal seams that are amenable to successful hydrofracking (refer to Chapter 5: Pore Pressure and Stress Field in Coal Reservoirs), a vertical borehole (well) is drilled to a point about 200 ft below the target coal seam and a steel casing is set just above the top of the coal seam. The coal seam is cleaned with high-pressure water (called hydrojetting) to remove any cement or debris. Next, a small amount of water is pumped into the coal formation (called minifrac) and the well is shut in to observe pressure decline. The data is used to design the main fracking (to be discussed later in the chapter). A precalculated amount of water mixed with good-quality sand is pumped into the coal formation. Under good conditions, the coal seam is split vertically from top to bottom and the fracture extends bilaterally to 500–1000 ft. After the designed volumes of water and sand are pumped, the borehole is kept shut for a few hours. After that, the well is flowed back slowly. Some gas production is noted, but the well is still full of water. A pump is installed on the well to pump out water. As the water depletes, gas production increases. Ultimately, all the water is pumped out (in about

6 months to 1 year) and only sand is left in the fracture, keeping the fracture open for gas production. The gas production may last 5–20 years depending on the size of the area being drained.

The subject matter will be discussed under the following headings:

1. Theoretical estimation of fracture dimensions; length, width, and height.
2. The direction of the fracture.
3. The fracture procedure:
   a. Water fracture of a vertical well.
   b. Foam fracture of a vertical well.
   c. Fracking of deep horizontal wells
4. Fracture pressure analysis.
5. In-mine mapping of fractures.

8.2 THEORETICAL ESTIMATION OF FRACTURE DIMENSIONS

The literature is replete with information on the subject, but it mostly deals with fracture geometry in sandstone, limestone, and shale formations. A coal seam is a very different reservoir, as discussed in Chapter 1, Global Reserves of Coal Bed Methane and Prominent Coal Basins.

When a given volume, V, of fluid and sand is pumped into a coal formation, it creates a fracture of volume V_F (mostly full of sand) and the remaining fluid, V_L, is lost in a process called "leak off." Conservation of mass requires that:

$$V = V_F + V_L \tag{8.1}$$

$$\text{Now } V = Q \times t$$

where Q is the pumping rate in ft^3/min
t is the time in minutes

$$V_F = L \times W \times H$$

L is the length of the fracture
W is the width of the fracture
H is the height of the fracture

$$V_L \text{ is equal to } (3\, H_P C L)\sqrt{t} \tag{8.2}$$

where
H_P is the wetted height of the fracture, usually equal to H.
C is the leak-off coefficient, a characteristic of the coal formation and fluid.

L is the length of the fracture, tip-to-tip.
t is the time.

8.2.1 Estimation of the Length of the Fracture

Eq. (8.1) can be rewritten as

$$Q \times t = L \times [3CH_P\sqrt{t} + WH]$$

$$\text{or } L = \left[\frac{Qt}{3CH_P\sqrt{t} + WH}\right] \tag{8.3}$$

where
L is the tip-to-tip length of the fracture.
Q is in ft^3/min [$1\ ft^3 = 5.615$ barrels].
t is time in minutes.
C has a dimension of $ft/\sqrt{min}$.
H_P, W and H are in feet.

In hydrofracking coal seams, it was observed that mostly $H = H_P$ and width is generally 0.5 inch to 1 inch (0.04–0.08 ft). An average of 0.06 ft can be assumed.

Hence Eq. (8.3) becomes

$$L = \frac{1}{H}\left[\frac{Qt}{3\ C\sqrt{t} + 0.06}\right] \tag{8.4}$$

It can be discerned from Eq. (8.4) that if we use low-viscosity fluid, such as water, to fracture, the H will be low, and correspondingly, one would get a larger L. In thin coal seams, it is desirable to get a large L and small H.

For a fixed Q and t and neglecting the width (a small quantity), Eq. (8.4) can be written in logarithmic form as

$$\log L = \log\left(\frac{Qt}{H}\right) - \log 3C - \frac{1}{2}\log t \tag{8.5}$$

Hence a plot of log L versus log C will yield a straight line, indicating that the higher the leak-off coefficient, the shorter the fracture.

Coal seams have, typically, a much higher leak-off coefficient, on the order of 0.01 to 0.05 $ft/\sqrt{min}$ compared to sandstone, which may have a value of 0.001 to 0.005 $ft/\sqrt{min}$.

Another way to express the leaked-off volume of fluid is in terms of efficiency, η,

where

$$\eta = \frac{V_F}{V} \tag{8.6}$$

Hence, total leak-off volume $V_L = V\ (1-\eta)$. This η should not be confused with the frac efficiency used to design hydrofracking of rock. The fracture efficiency in coal is generally low.

An example:

In hydrofracking coal, 120,000 gallons of water with 100,000 pounds of sand was used to create a fracture 2000 ft long, 10 ft high (average) and ½ inch wide (average). Calculate the efficiency of fracture:

Here,

$$V = 16,604.7\ \text{ft}^3$$
$$V_F = 8333.3\ \text{ft}^3$$

Hence, $\eta = 5\%$.

We can also calculate the leak-off coefficient from these data. Assuming a rate of pumping at 30 bbl/min, the pumping time is 95.23 minutes.

Putting these values into Eq. (8.2), we get a value of $C = 0.0269$. This is at least one order of magnitude higher than the value for sandstone.

8.2.2 Fracture Width Estimation

When the gas well in a coal seam is pressurized to fracture it, the coal seam opens up with an elliptical cross section (Fig. 8.1) with a maximum width, $W_{max.}$

Where

$$W_{max} = \frac{2h\,(P-\sigma)}{E^1} \tag{8.7}$$

where

h = Height of coal
P = Bottom-hole pressure when fracture opens up
σ = Reservoir gas pressure

$$E^1 = \frac{E}{(1\text{-}\nu^2)} = \text{plane strain modulus}$$

E = Young's modulus for coal
ν = Poisson ratio for coal

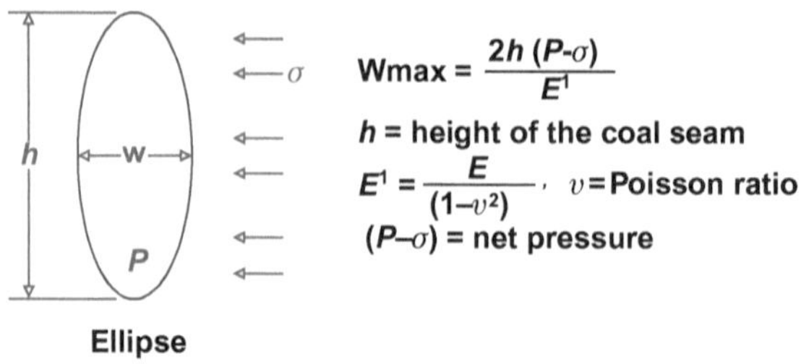

Figure 8.1 Fracture opening in the coal seam.

An example:

A 5-foot thick coal seam fractures at 3000 psi bottom-hole pressure and has a reservoir pressure of 500 psi. Calculate the maximum width of fracture assuming

$$E = 500,000 \text{ psi}$$
$$v = 0.3$$

$$\text{Calculate } E^1 = \frac{500,000}{1 - 0.09} = 549,451 \text{ psi}$$

Putting these values into Eq. (8.7)

$$W_{\max} = \frac{2 \times 5\ (3000 - 500)}{549,451} = 0.0455 \text{ ft or } 0.54 \text{ inches}$$

Several published papers by Nolte [1], Perkins and Kern [2], Geertsma and deKlerck [3] and Warpinski [4] discuss the fracture width calculation in great detail.

There are three main models:

1. The Perkins and Kern (P-K) Model:

   The fracture is assumed to be elliptical in cross section and it gets narrower as the fracture extends to the tip bilaterally, as shown in Fig. 8.2. The fracture stops extending when $P = \sigma$ in Eq. (8.7). This model is most applicable for water fractures in coal, where our observations indicate $L >> H$. The height may be only 20 ft for a fracture that is 2000 ft long, tip-to-tip.

2. The Geertsma and deKlerck Model:

   The cross-section of the fracture is assumed to be rectangular, with sides equal to W and H. As the fracture extends to the tip, the width narrows, but the height of the fracture remains constant. This model is applicable only where L is only slightly larger or even smaller than H.

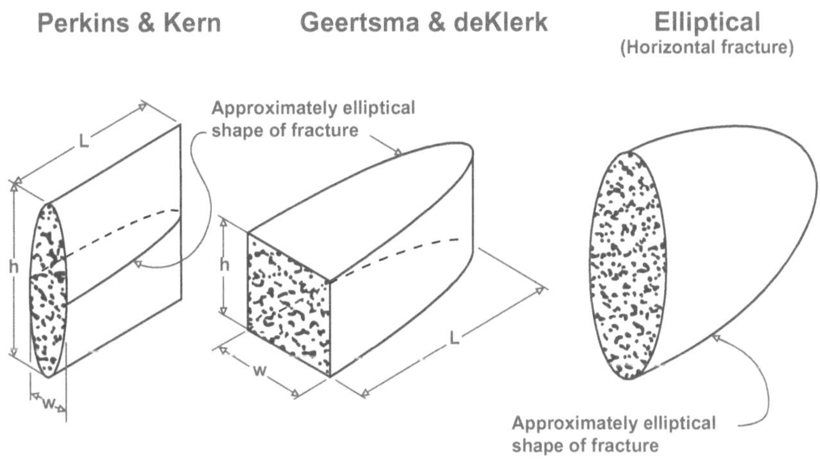

Figure 8.2 Theoretical concepts of fracture geometry.

It is more applicable to a fracture created by a very viscous fluid, such as cross-linked borate gel or heavy nitrogen foam.

3. The Radial/Elliptical Model:

In this case, the fracture opens up as an ellipse, but extends with a mean radius R or as an ellipse. It is mostly applicable to a horizontal fracture when the coal seam does not fracture, but the junction at the roof fractures, depositing sand in a circular or elliptical pattern. This usually happens in a shallow well when the vertical stress, $\sigma_{v,}$ is the lowest stress.

Using the Perkins and Kern model, one can derive another expression for w which shows its dependence on the rate of flow of frac fluid, its viscosity, and the plane strain modulus of the formation.

Based on field observation,

$$W_{\text{average}} = \frac{\pi}{4} W_{\max}$$
$$W_{\text{average}} = \frac{\pi}{4} \times \frac{2H(P-\sigma)}{E^1} \tag{8.8}$$

From the equation for fluid flow in a narrow strip, given by Craft and Hawkins [5],

$$\frac{P-\sigma}{L} = \frac{12\mu Q}{HW^3} \tag{8.9}$$

Substituting Eq. (8.9) into Eq. (8.8) for P-σ,

$$W = \frac{\pi}{4} \times \frac{2HL}{E^1}\left(\frac{12\mu Q}{HW^3}\right)$$

$$\text{or } W^4 = 6\pi\left(\frac{\mu QL}{E^1}\right) \quad (8.10)$$

$$\text{or } W = 2.1\left(\frac{\mu QL}{E^1}\right)^{\frac{1}{4}}$$

This shows that average width is proportional to the 1/4 power of the flow rate and proportional to $(E^1)^{1/4}$.

An example:

Calculate the average width of a fracture where

$L = 2000$ ft

$Q = 30$ bbl/min $= 2.8$ ft^3/sec

$E^1 = 550{,}000$ psi

$$\mu = 1cp = 2,08 \times 10^{-5}\ lb\text{-}sec/ft^2$$

Plugging these values into Eq. (8.10),

$W = 0.54$ inch.

The agreement with Eq. (8.7) is remarkable.

Corollaries:

a. If the frac is done at a fluid flow rate of 60 bbl/min, the width will increase by $(2)^{1/4}$ or 1.19 times.

b. If the viscosity of the fluid is increased by 1000 times, the width of the fracture will increase 5.6 times.

c. If E^1 for the roof increases by 10 times (sandstone), the fracture width will reduce by a factor of 1.8. Typical fracture width in shale roofs is 1/8–1/4 inch.

8.2.3 The Height of the Fracture

A successful hydrofracture of a coal seam aims to create a vertical fracture from the bottom of the coal seam to the top. The fracture almost never penetrates the floor of the coal seam (as one would expect from the laws of fluid dynamics—fluid flows down the pressure gradient), but it can go up, beyond the top of coal into the roof. The total height of the fracture at the well bore is controlled by the net pressure, $(P - \sigma)$, and the mechanical properties of the roof rock, such as its modulus of elasticity and compressive strength. The higher the modulus of elasticity and compressive strength, the smaller the height growth and the width of the fracture. A soft coal with a

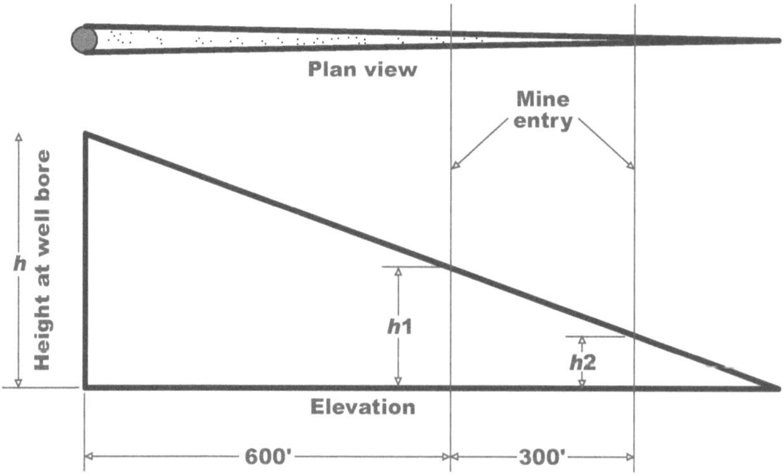

Figure 8.3 Calculation of fracture height by similar triangles.

compressive strength of 3000 psi and E of 500,000 psi was overlain by a strong marine shale with a compressive strength of 13,000 psi and an E of 3×10^6 psi. After hydrofracking and mining out of the area, the total fracture height was only 20 ft at the well bore. The width of fracture in coal was ½–¾ of an inch, but the width in the roof shale was about ⅛ of an inch at the top of the coal and only a trace 15 ft above the coal.

The ground stress in coal and shale did not appear to have any impact on height. As explained in Chapter 5, Pore Pressure and Stress Field in Coal Reservoirs, no significant difference in the ground stresses in coal and shale is anticipated. The fracture height at the well bore was calculated as shown in Fig. 8.3.

The fracture was cut by mine roadways at 600 and 900 ft from the well bore. The height of the fracture from the ground was measured as h_1 and h_2. By using similar triangle analysis, fracture height at the well bore, h, was calculated to be equal to $3\,h_1 - 2\,h_2$.

Thus, if h_1 was 9 ft and h_2 was 6 ft, the total fracture height at the well bore is estimated at 15 ft. This was a water fracture that normally follows the P-K model where $L >> H$.

8.2.4 The Direction of the Fracture

After mining through more than 200 wells, the following theories about hydrofracking were confirmed.

1. The fracture volume ($L \times W \times H$) is proportional to the total fluid volume.

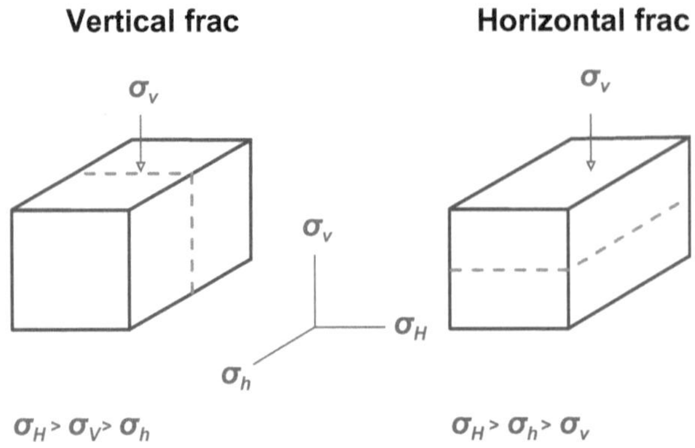

Figure 8.4 Dependence of fracture direction on ground stress.

2. Coal has a higher leak-off coefficient than sandstone and other rocks (by one order of magnitude).
3. Most often, the fracture does not travel in the floor.
4. The direction of fracture is always orthogonal to the least stress.
5. That is, if $\sigma_H > \sigma_v > \sigma_h$, the fracture would be vertical.
6. But, if $\sigma_H > \sigma_h > \sigma_v$, the fracture would be horizontal (Fig. 8.4). Most often, the coal separates from the roof and all the sand is deposited there in an elliptical area. The major axis of the sand deposit is in the direction of σ_H, and the minor axis is parallel to σ_h.
7. The width of the fracture is proportional to $(w)^{1/4}$ and $(Q)^{1/4}$, but it is mainly controlled by the elastic modulus of the formation as shown in Eq. (8.10).
8. In a composite formation of coal, shale, and sandstone, the horizon with the lower elastic modulus will fracture first. The Poisson ratio does not seem to have any influence.

Table 8.1 shows the actual length, width, height, and azimuth of 10 water fractures in coal where the direction of σ_H was N 55°E as measured in the mine. A typical hydrofrac job used 120,000 gallons of water, 100,000 pounds of sand (mostly 20–40 mesh), and was fracked at 30 bbl/min. Details are given in Section 8.3.

The remaining values are estimated.

8.3 THE FRACTURING PROCEDURE

Fig. 8.5 shows a typical vertical cross section of hydrofracked well in a 5–1/2 foot thick coal seam.

Table 8.1 Dimensions and directions of water frac in coal

Well number	Total length (ft)[a]	Width at well bore (inches)	Height (ft)	Avg. direction (left and right wings)
1	1875	0.75	20[b]	N58–60E
2	1750	0.5	20	N54–59E
3	1450	0.5	20	N50–53E
4	1600	0.5	20	N57–57E
5	1750	0.5	20	N53–60E
6	1625	0.5	20	N53–54E
7	1750	0.75	20	N57–57E
8	1550	0.75	20	N57–57E
9	1650	0.75	20	N55–57E
10	2100	0.75	20[b]	N55–57E
AVG	**1550**	**0.6**	**20**	**N56E**

[a]The propped length was typically 75% of the total length.
[b]Actually measured by cutting the roof.

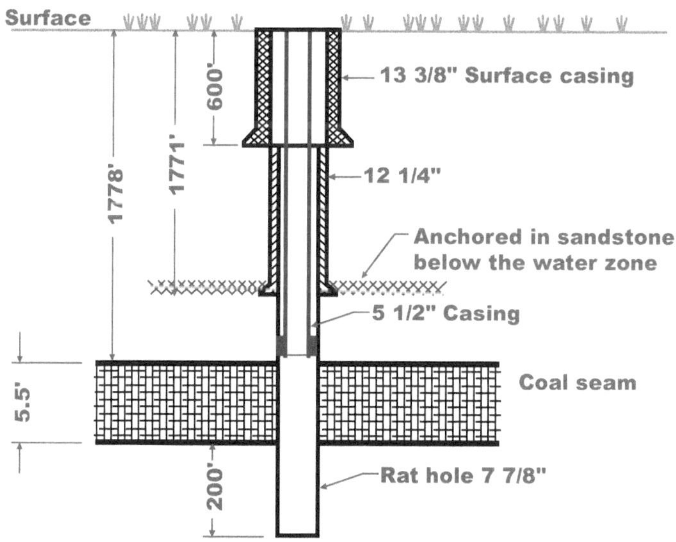

Figure 8.5 A typical well completion in a single coal seam.

Procedure:

1. A 15-inch diameter borehole is drilled to 600 ft depth and a 13–3/8-inch casing is set.
2. A 12–1/4-inch diameter borehole is drilled to 1771 ft, and a 9–5/8-inch casing is set.

3. Next, a 7–7/8-inch borehole is drilled to the target depth, which is typically 150–200 ft deeper than the target coal seam. This portion of the borehole is called a "rat hole," and it collects coal and sand particles that fall out of the fracture. The borehole is routinely cleaned with a bailer during gas production.
4. A formation packer is used to cement the 5–1/2-inch casing at a good spot just above the target coal seam. The rat hole is filled with sand to the floor of the coal seam. The well is tested to see if there is any water leakage from the packer at 3000–4000 psi. The pipelines are tested at 6000 psi.
5. Production horizon: 1778–1785.5 ft

   Total Depth: (1785.5 + 200) = 1985.5 ft

First Day's Work:

6. The next step is to hydrojet the coal seam with 2600 psi water at 4 bpm for 20 minutes to remove any cement that may have leaked into the coal seam.
7. Next, a minifrac is done to estimate reservoir properties. All details are shown in Table 8.2. The data is used to calculate many important fracture parameters as discussed later.
8. The layout of all hydrofracking equipment is checked. The following equipment is needed:
   - a. Two HT 400 pumps capable of giving a combined 40 bpm.
   - b. One blender with a capacity of 50 bpm.
   - c. Several sand trucks to deliver
     1. 15,000 pounds of 80–100 mesh sand
     2. 100,000 pounds of 20–40 mesh sand
     3. 15,000 pounds of 10–20 mesh sand
   - d. 3000 bbl of water in twelve 250 bbl tanks
   - e. One hydrofracking van

   The well was shut-in for hydrofracking the next day.

Next Day

9. Work is started at first light so there will be time to handle any mechanical, hydraulic, or sand-screening problem.
10. A safety meeting is held. All personnel are counted and radio communications is checked. No person is in a direct line from the well head.
11. The hydrofracking starts. All details are shown in Table 8.3 (stage 1) and Table 8.4 (stage 2).
12. The well is shut-in for all equipment (called irons) to move out.

Table 8.2 Minifracture data

Flow rate (bbl/min)	Volume (bbl)	Cum. volume	Cum. time (minutes)	Pressure (psi)
1	10	10	10	900–975
2	10	20	15	1150–1200
6	15	35	17.5	1400
10.5	35	70	21	1600–1520

Pumping stopped and the well was shut in. ISIP (Instantaneous Shut-in Pressure) was 1250 psi.

$$\text{Frac Gradient} = \frac{\text{ISIP}}{\text{Depth of Borehole}} + 0.434 = \frac{1250}{1780} + 0.434$$
$$= 0.702 + 0.434$$
$$= 1.136$$

The frac gradient (FG) is low. The fracture may have communicated with a previous well.

Pressure decline after shut-in:

Time, t (min)	Pressure (psi)
1	1100
2	1050
5	900
10	750

13. After 4 hours, the well is flowed back through either a choke or an open-hole.

14. In 3–4 hours the well ceases to produce any water or sand. Some gas flow is usually detected.

15. All personnel are accounted for, and the well is handed over to the production crew for swabbing and installation of a water pump.

8.4 FOAM FRAC FOR COMMERCIAL GAS PRODUCTION

Hydrofracking with slick water (water mixed with a friction-reducing compound) is ideal for coal seam degasification because it creates a long fracture ($L \gg H$) in the coal seam. For commercial gas production, it becomes necessary to hydrofracture all coal seams that are amenable to hydrofracking in one well. Assume three to four horizons, each containing five- to six-foot thick coal seams, are to be hydrofracked. It becomes

Table 8.3 Stage one of hydrofracture

Step	Volume pumped (bbl)	Cumulative volume (bbl)	Rate	Sand	Surface pressure	BHP
1 Pad	400	400	36	0	1450	1861
2	25	425	36	X^{3a}	1340	
3	25	450	35	0	1355	
4	25	475	35	X^3	1370	
5	25	500	35	0	1350	
6	50	550	35	Y^2	1394	
7	50	660	35	0	1400	1866
8	50	650	35	Y^3	1338	
9	50	700	35	0	1338	
10	75	775	35	Y^3	1470	
11	75	850	35	0	1450	2000
12	50	900	34	Y^3	1530	
13	50	950	35	0	1400	2036
14	50	1000	35	Y^3	1490	
15	50	1050	35	0	1430	
16	50	1100	35	Y^3	1415	
17	50	1150	34	0	1467	
18	50	1200	33	Y^3	1610	
19	50	1250	34	0	1450	
20	50	1300	34	Y^3	1550	
21	50	1350	34	0	1470	2150
22	50	1400	33	Y^1	1600	
23	50	1450	32	0	1530	2167
24	50	1500	32	$Y^{1.25}$	1580	
25 Flush	150	1650	32	0	1530	1991
26	ISIP 1065 psi; at 1 min: 952 psi; 2 min: 906 psi; 3 min: 874 psi; 5 min: 850 psi; 9 min: 814 psi; 10 min: 777 psi; 15 min: 735 psi; 20 min: 713 psi; 30 min: 645 psi (reservoir pressure).					

X is 80–100 mesh sand.
Y is 20–40 mesh sand.
Z is 10–20 mesh sand.
[a]The exponent "3" means at 3 lb/gallon.

difficult to procure and store 500,000 gallons of water on a small location. The preferred course is to change the frac fluid and use nitrogen foam.

The goal is to stimulate all coal seams in the zone isolated for fracturing in a single operation, but seams less than two feet in thickness are generally not stimulated. Nitrogen foam has a viscosity in the range 100–150 cp. It will create a wider fracture with enough height growth to intersect all the coal seams in the production horizon.

Table 8.4 Stage two of hydrofracture

Step	Volume pumped (bbl)	Cumulative volume (bbl)	Rate (bbl/min)	Sand	Surface pressure PSI	BHP PSI
1 (Pad)	300	300	38	0	1580	1950
2	50	350	35	X^3	1520	
3	50	400	34	0	1540	
4	50	450	34	Y^3	1550	
5	50	500	34	0	1450	
6	50	550	34	Y^3	1570	
7	50	600	34	0	1490	
8	50	650	34	Y^3	1630	
9	50	700	34	0	1480	
10	50	750	33	Y^3	1670	
11	50	800	34	0	1500	
12	50	850	34	Y^3	1660	
13	50	900	33	0	1615	
14	50	950	33	Z^1	1750	
15	25	975	33	0	1600	
16	50	1025	32	$Z^{1.5}$	1580	
17	50	1075	33	0	1580	
18 Flush	100	1175	34	0	1530	
	ISIP 1204 PSI					

Summary:

Stage	Volume (bbl)	Pressure Avg	Rate	Sand (lbs)			ISIP	FG	BHP
				X	*Y*	*Z*			
1	1650	1500	35	6.3k	51k	4.7k	1065	1.03	2036
2	1175	1600	34	6.3k	29k	5.25k	1204	1.11	2089
Total	2825	–	–	12.6k	80k	9.95k	–	–	–

This created a fracture of 1400 ft that was 0.75 inch wide at the well bore and about 20 ft high. Total gas production in 1000 days was 70 MMCF. The first year average production was 111 MCFD.

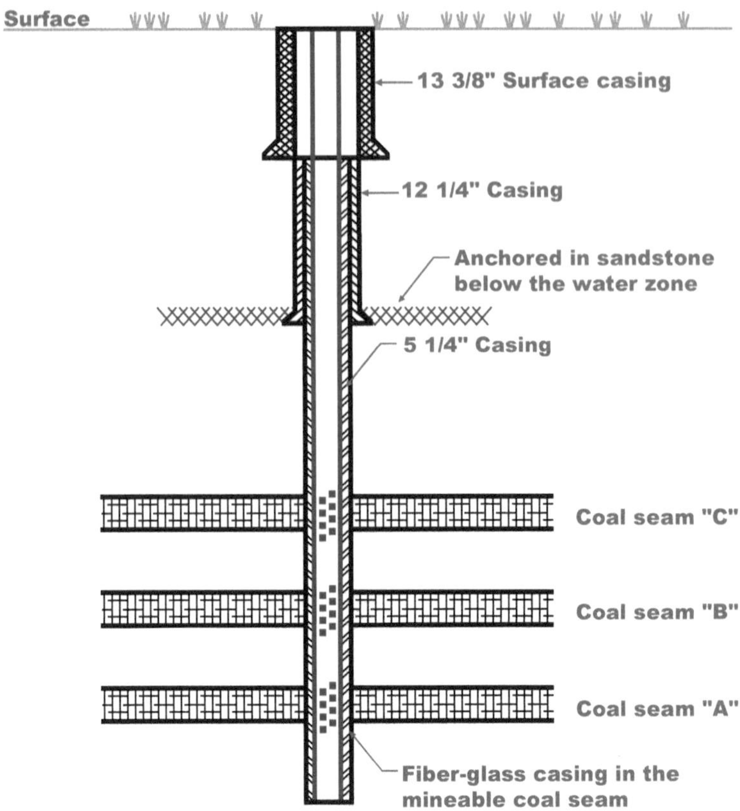

Figure 8.6 A typical multiseam well completion for foam fracture.

The nitrogen foam frac fluid has the following composition:

Nitrogen: 70% by volume

Gel: 15–20 lbs/1000 gallons, soluble in water

Water: 30% by volume

A foaming compound, such as SS0–21 (a commercial product)

The fluid is called 70% (Nitrogen) foam and has a viscosity of 100–150 cp. It can carry 4–6 pounds of sand per gallon, but the concentration is normally kept below 4 lbs/gallon to avoid sand-screening.

A typical multifrac well is shown in Fig. 8.6.

Three horizons containing many coal seams are stimulated. It does not pay to stimulate seams less than two feet thick.

Table 8.5 shows the details of a typical well with three production horizons.

The rate of pumping was 35–36 bbl/min. Because of the higher viscosity (100 cp), the fractures are wider. Eq. (8.10) predicts a width of

Table 8.5 A typical three-stage foam frac

Stage	Frac horizon (ft)	Coal thickness (ft)	Nitrogen (MCF)	Water (gal)	Gel (lb/1000 gal)	Foamer SS0-21	Sand (100 lbs)	F.G.
1	1560–1565	5	553	24,990	20	SSO-21	793	1.5
2	1381–1390	7	442	16,674	20	SSO-21	505	1.55
3	914–1192	9	472	18,774	20	SSO-21	615	1.57
Total		21	1467	60,438			1913	

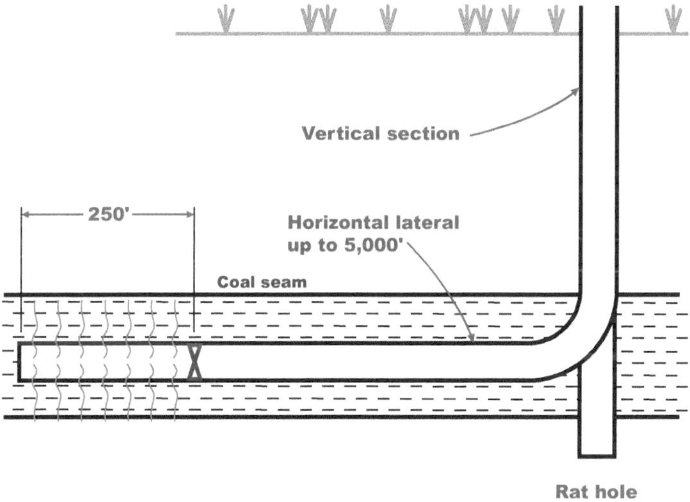

Figure 8.7 A typical completion in a deep horizontal well.

(0.5–0.75) $(100)^{1/4}$ or 1.6–2.4 inches. The observed width at the well bore was about 2 inches.

The length of the fracture is correspondingly reduced. Assuming a frac height of 30 ft and a pumping time of 30 minutes for Stage I, L is calculated using Eq. (8.3) and is equal to 336 ft. The observed length was approximately 360 ft.

8.5 SLICK WATER HYDROFRACKING OF HORIZONTAL WELLS IN DEEP FORMATION

A recent development in hydrofracturing of deep shale formations is applicable to deep coal seams all over the world.

In a typical case, horizontal boreholes of 3000–5000 ft length are drilled bilaterally in the formation as shown in Fig. 8.7. The horizontal leg is hydrofracked every 250 ft through multiple perforations in the casing using slick water (water mixed with a friction-reducing compound).

It takes nearly 500,000 gallons of water with 600,000 pounds of sand to complete one stage of 250 ft length. The pumping schedule is shown in Table 8.6. If both legs are fracked at every 250 ft, it will take 40 stages to do it, consuming 20 million gallons of water and 24 million pounds of sand. This is a massive job costing millions of dollars. In Marcellus shale, a production of 5–10 MMCFD is realized leading to a net profit. The well life is usually 20 years. The Marcellus shale gas content is only 75 ft^3/t while the deep coal seams contain 400–600 ft^3/t. Coal seams with a thickness of 40–60 ft can produce 10–20 MMCFD if properly hydrofracked and equally carefully produced. As discussed in Chapter 1, Global Reserves of Coal Bed Methane and Prominent Coal Basins, the bulk of the coal deposits are deeper than 3000 ft and are very amenable to commercial gas production.

8.6 FRACTURE PRESSURE ANALYSIS

Hydrofracking of a coal seam in a working mine has the unique advantage that the fracture is eventually mined out and the length, width at the well bore, and direction can be directly measured. The fracture is cut in several places away from the well bore where the height can be directly measured. The height at the well bore is calculated as discussed earlier. When hydrofracking a formation that will not be mined, the growth of the fracture can be estimated by fracture pressure analysis.

During the hydrofracturing process, the bottom hole pressure (bhp) is monitored continuously and for a short time after the pumping is terminated. Fig. 8.8 shows a simplified recording of the bhp in a vertical well. Since most fractures (water or foam) in coal have long lengths compared to the height of the fracture, the P-K model is the most applicable. For the P-K model, Eq. (8.11) defines the fracturing pressure.

$$P_{\text{net}} = (P - \sigma) \simeq \frac{E^1}{H} (\mu L Q)^{\frac{1}{2n+2}} \tag{8.11}$$

where

P_b = bottom hole pressure
σ = reservoir pressure
n is a characteristic of the frac fluid
$n = 1$ for water or slick water
$n = 0.5$ for nitrogen foam

Table 8.6 A typical slick water fracture schedule in marcellus shale

Stage	Proppant type	Start BH prop conc (lb/gal)	End BH prop conc (lb/gal)	Clean volume (gal)	Start clean rate (bbl/min)	Start slurry rate (bbl/min)	Start blender prop conc (lb/gal)	Prop mass (lb)	Cumulative prop mass (lb)	Stage time (min)
Acid				3000	15.0	15.0				4.8
Pad				4200	30.0	30.0				3.3
Acid				3000	15.0	15.0				4.8
Pad				35,000	85.0	85.0				9.8
Proppant-laden fluid	100 mesh	0.25	0.25	32,000	84.0	85.0	0.25	8000	8000	9.1
Proppant-laden fluid	100 mesh	0.50	0.5	42,000	83.1	85.0	0.50	21,000	29,000	12.0
Proppant-laden fluid	100 mesh	1.00	1	59,000	81.3	85.0	1.00	59,000	88,000	17.3
Proppant-laden fluid	100 mesh	1.50	1.5	62,000	79.6	85.0	1.50	93,000	181,000	18.6
Proppant-laden fluid	40/70 Premium White	1.00	1	20.000	81.3	85.0	1.00	20,000	201,000	5.9
Proppant-laden fluid	40/70 Premium White	1.50	1.5	17,000	79.6	85.0	1.50	25,500	226,500	5.1
Proppant-laden fluid	30–50 Premium White	1.00	1	39,000	81.3	85.0	1.00	39,000	265,500	11.4
Proppant-laden fluid	30–50 Premium White	1.50	1.5	39,000	79.6	85.0	1.50	58,500	324,000	11.7
Proppant-laden fluid	30–50 Premium White	2.00	2	88,000	77.9	85.0	2.00	176,000	500,000	26.9
Proppant-laden fluid	30–50Premium White	2.50	2.5	40,000	76.3	85.0	2.50	100,000	600,000	12.5
Flush				11,500	85.0	85.0			600,000	3.2
Total				**494,700**				**600,000**		**156.2**

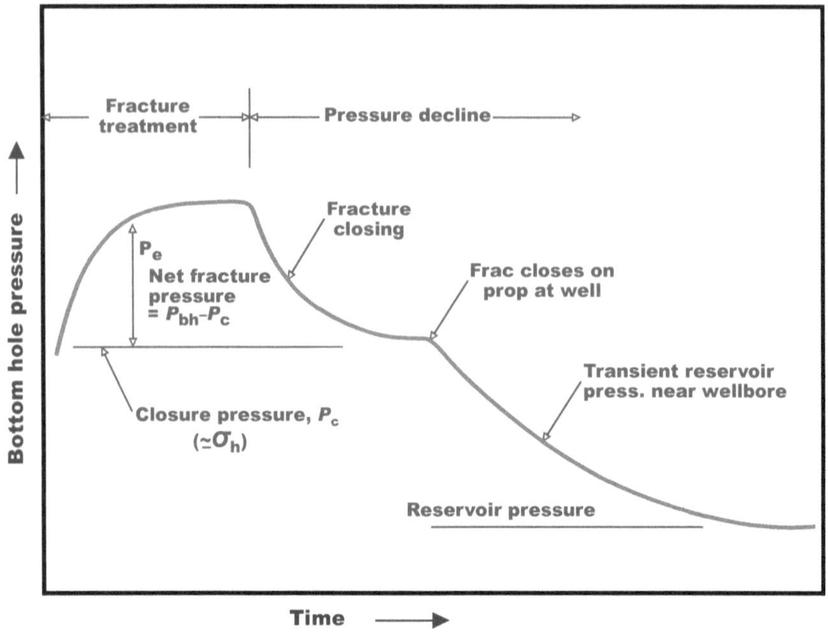

Figure 8.8 A typical profile of bhp versus time in a hydrofracked well.

Other terms are the same as defined earlier.

Eq. (8.11) thus changes into

$$P_{\text{net}} = \frac{E^1}{H}(\mu LQ)^{1/4} \text{ for water frac and}$$

$$P_{\text{net}} = \frac{E^1}{H}(\mu LQ)^{1/3} \text{ for nitrogen foam frac}$$

From Fig. 8.8, four different phases in the pressure profile can be identified. As pumping starts, the bhp builds up and the formation opens up to receive the fracking fluid. The net pressure is basically dependent on the length of the fracture, because all other parameters in Eq. (8.11) are constant. However, the length depends on the volume pumped (Qt) or just time, t, because Q is generally constant.

Nolte and Smith (1979) plotted Fig. 8.8 on a log–log scale to clearly show the four modes in the fracturing processes as straight lines.

Fig. 8.9 shows the four modes of the fracture process.

The characteristics of each mode are described below.

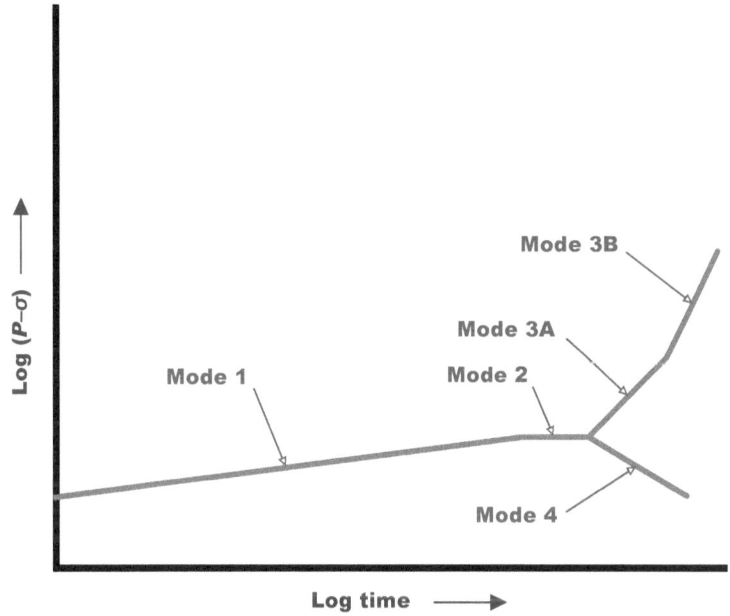

Figure 8.9 Log–log plot of effective fracture pressure against time.

Mode	Slope of the line	Interpretation
I	1/8–1/4	Unrestricted linear extension of the fracture; restricted height
II	0 (straight line)	Moderate height growth; fracture extension continues
III	1 in 1 (45 degrees) or	Restricted extension: two wings
	2 in 1 (63.4 degrees)	Restricted extension: one wing
IV	Negative gradient	Unstable height growth

In coal, the fracture has maximum height and width at the well bore. As the fracture extends and becomes longer, both the width and height decrease. The extension of the frac stops either when $P_{net} = \sigma$ or when all fluid is lost in the formation due to high leak-off. Mode IV is generally not seen in coal seam fracking.

Further discussion of modes I, II, and III are provided below.

Mode I: The straight line with a slope of 1/8–1/4 indicates that the fracture is propagating linearly with confined height. The injection rate and fluid viscosity (amount of sand in the slurry) can remain constant. The fracture is extending as predicted by the P-K model.

Mode II: A flat pressure line indicates stable height growth or increased fluid loss that stops any increase in pressure. In coal fracturing, the latter is true because the height of the fracture declines the farther the fracture extends from the well bore. Eventually, a stage is reached where the slurry is dehydrated and a sand bridge is formed in the fracture. This leads us to mode III.

Mode III: This mode is characterized by a positive slope (45–63 degrees on a log–log plot), indicating flow restriction where incremental volumes pumped correspond to incremental pressure increase. The approximate distance from the well bore where this sand bridge occurs, L_{max}, is given by

$$L_{max} = \frac{1.8\ QE^1}{h^2(\Delta P/\Delta t)} \tag{8.12}$$

assuming

$Q = 30$ bbl/min

$E^1 = 500{,}000$ psi

$h = 5$ ft

$\Delta P/\Delta t = 1000$ psi/min (in Mode III)

$$L_{max} = \frac{1.8 \times 30 \times 500{,}000}{25(1000)} = 1080 \text{ ft}$$

8.7 ANALYSIS OF MINIFRAC DATA

A lot of information about the reservoir can be obtained by some simple analysis of the minifrac data.

8.7.1 Instantaneous Shut-in Pressure (ISIP)

As soon as the calculated volume of the fluid is pumped in, the well is shut in. The instantaneous bhp is recorded and called ISIP. It is used to calculate the FG of the well.

$$\text{F.G.} = \frac{\text{ISIP}}{\text{Depth}} + \frac{\text{Hydrostatic Head}}{\text{Depth}}$$

Fig. 8.10 shows the plot of bhp against time.

ISIP = 1250 psi

Depth of the well: 1781 ft (middle of the coal seam).

Hence

$$\text{F.G.} = \frac{1250}{1781} + 0.434 = 1.14$$

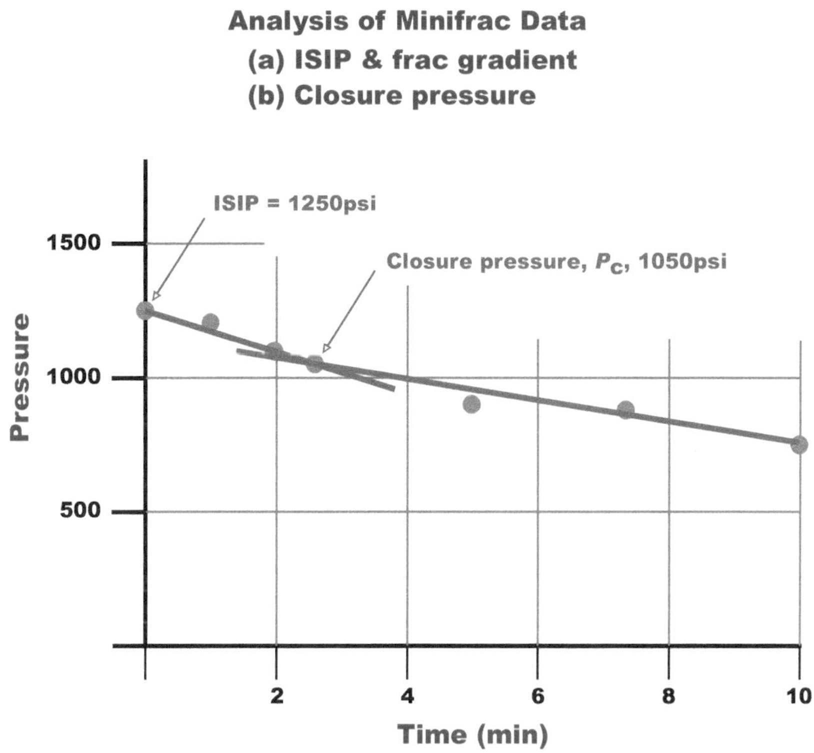

Figure 8.10 A plot of bhp against time.

This shows a very high permeability. Most likely, the well communicated with another previously fractured well. The pressure declines at a given gradient, but after a few minutes, when it equals the lowest horizontal stress (σ_h), the slope changes. This particular pressure point is called closure pressure or P_c as shown in Fig. 8.10.

$P_c = 1050$ psi. It is roughly equal to σ_h (the minor horizontal stress).

8.7.2 Horner's Plot for Reservoir Pressure

The data from minifrac can be again plotted as

$$\text{bhp vs } \log\left(\frac{t_o + t_{si}}{t_{si}}\right)$$

where

t_o is the time of pumping and

t_{si} is the time interval when pressure readings were taken.

Fig. 8.11 shows a plot. The point where the straight line cuts into the y axis is the reservoir pressure, P^*. The data is presented in Table 8.7. This

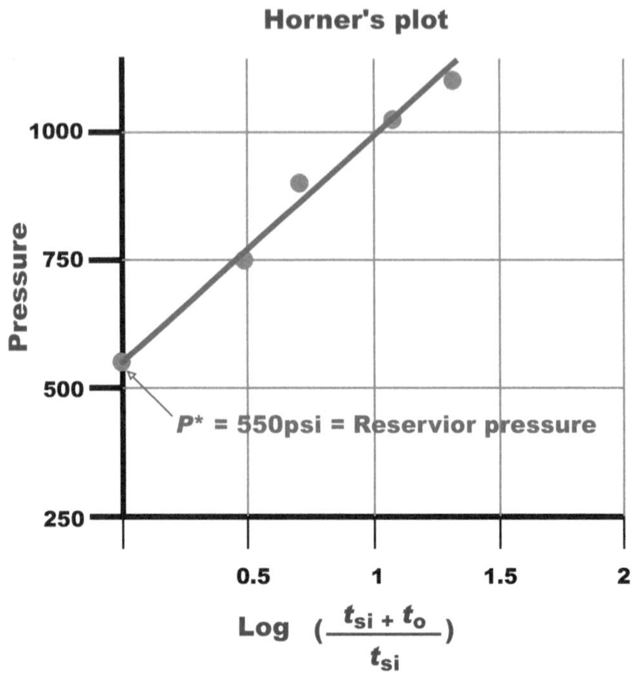

Figure 8.11 Horner's plot of minifrac data.

Table 8.7 Calculation of minifrac data for Horner's plot

$t_o = 21$ min

t_{si} (min)	$\frac{t_0 + t_{si}}{t_{si}}$	Pressure (psi)	Log $\left(\frac{t_{si} + t_o}{t_s}\right)$
1	22	1100	1.34
2	11.5	1050	1.06
5	5.2	960	0.71
10	3.1	750	0.49

graph is known as Horner's plot. The reservoir pressure is 550 psi, which agrees with field measurements.

8.7.3 The Fracture Extension Pressure

The net fracture extension pressure can be obtained by plotting the bhp against the injection rate. The rising pressure gradient has an inflection point as shown in Fig. 8.12. This point is called the fracture extension pressure. For this well, it was 1275 psi. It is advisable that the bhp during hydrofracking should not exceed the hydrostatic head plus fracture extension pressure, which is 2047 psi in this case.

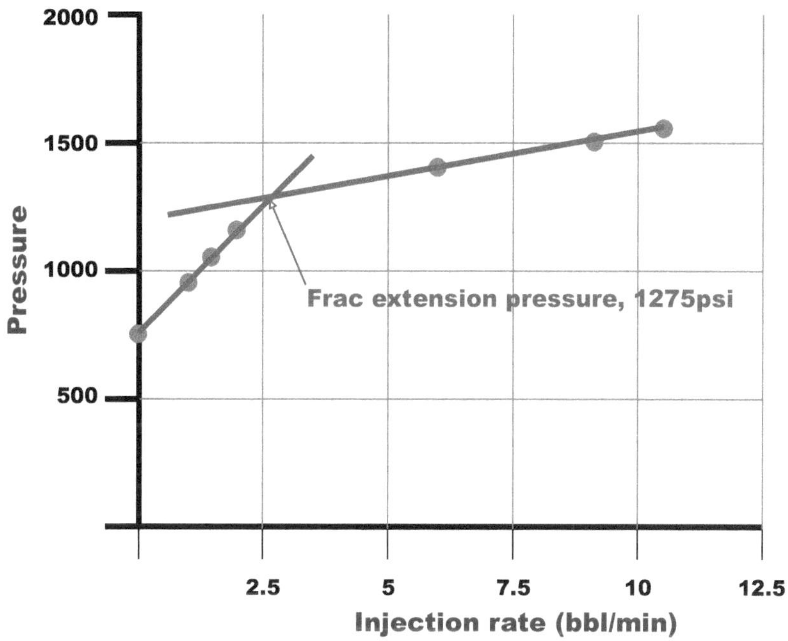

Figure 8.12 Plot of bhp against step-by-step injection rate.

8.8 MAPPING OF FRACTURES IN COAL MINES

Hydrofracking of coal in a working mine provides a unique opportunity to cut through the fractures and actually measure the length, width, height (in some cases), and direction of the fracture. The data was used to verify the theoretical estimates of the fracture dimensions and directions earlier. A detailed discussion is provided now.

It would be safe to say that no two fractures created by identical hydrofracking procedures are the same in shape or size. Maps of over 200 wells can, however, be grouped in three broad categories:

1. vertical fractures
2. horizontal fractures, and
3. mixed fractures.

8.8.1 Vertical Fractures

Fig. 8.13 shows the various types of vertical fractures. The necessary condition for a vertical fracture is $\sigma_H > \sigma_v > \sigma_h$. This is true for all types of fluids. A water fracture (at 30 bbl/min) typically creates a vertical fracture with a width of 0.5–0.75 inch at the well bore. The fracture does not extend into the floor (strong shale with a compressive strength

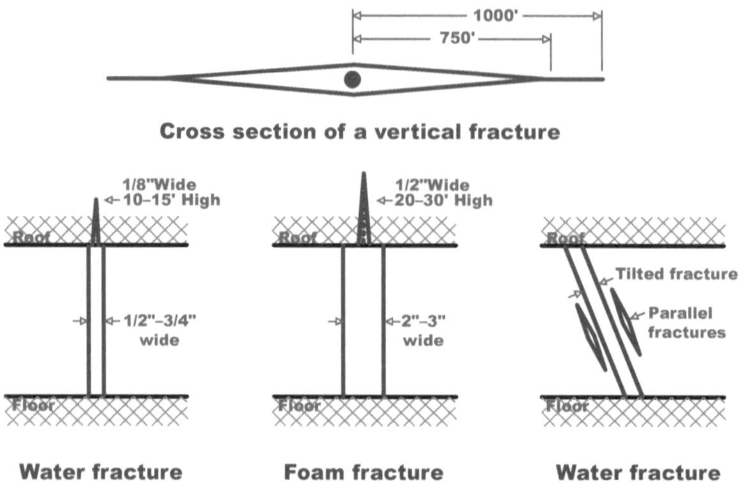

Figure 8.13 A typical vertical fracture.

of 13,000 psi), but does extend into the roof which is an identical rock to the floor. The fracture width is drastically reduced to about ⅛ inch and it can extend 10–15 ft into the roof. The fracture is about 2000 ft long tip-to-tip, but the propped length is typically 1500 ft (both wings combined).

The foam fracture (at the same rate) created a wider fracture of 2–3 inch width. It also extended higher into the roof, to 20–30 ft, as predicted by the theory. The length of the fracture was reduced to about 700 ft, about half the length created by water fracture. Many times, the fractures were not truly vertical but inclined at 30–45 degrees from the vertical with multiple, narrow parallel fractures that contained little sand. Most wells with a good vertical fracture had a FG of 1.3–1.5. They produced better than other wells, but this will be discussed in another section of this chapter.

8.8.2 Horizontal Fractures

Fig. 8.14 shows a typical horizontal fracture. The necessary condition is $\sigma_H > \sigma_h > \sigma_v$, i.e., the vertical stress is lowest in magnitude. This happens in shallow wells, typically less than 1500 ft in depth.

A water fracture in a shallow well mainly separates the coal seam from the roof and all sand is deposited there. The shape of the deposit is an ellipse with the longer axis parallel to σ_H (the face cleat in coal seams). The two axes of the ellipse are generally 250 × 100 ft. The thickness of

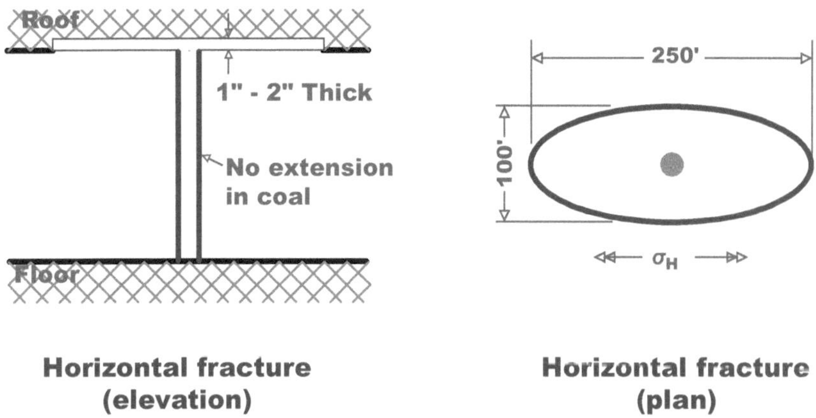

Figure 8.14 A typical horizontal fracture.

the sand deposit is 1–2 inches. Most wells with horizontal fractures had a FG of 1.6–2.0 and beyond. Such wells are poor gas producers. Hence, it does not generally pay to hydrofracture coal seams shallower than 1500 ft. In another basin, eight vertical wells were drilled in a 1000-foot deep coal seam and all were fractured with 70% nitrogen foam, but none of them produced any measurable quantity of gas.

8.8.3 Mixed Fractures

Gas wells in the depth range of 1500–1800 feet where σ_v is only slightly larger or smaller than σ_h resulted in mixed fractures. That is, part of the fracture was vertical and part of it was horizontal. Such fractures are called T fractures and are shown in Fig. 8.15.

The vertical fracture was as wide as the pure vertical fractures but much shorter. The horizontal fracture portion is also smaller than a pure horizontal fracture. A typical T fracture may have a total wing extension of only 100–300 ft.

The horizontal fracture again penetrates the roof (only 1 in 100 is in the middle of the coal seam), but its dimensions are smaller than a pure horizontal fracture.

8.8.4 Hydrofrac Wells With Low Frac Gradient

In the Central Appalachian Basin, when gas wells are drilled with small spacing (20–40 acres/well), it is likely that the fracture will run into a previously fractured well. This results in a FG of less than 1.3.

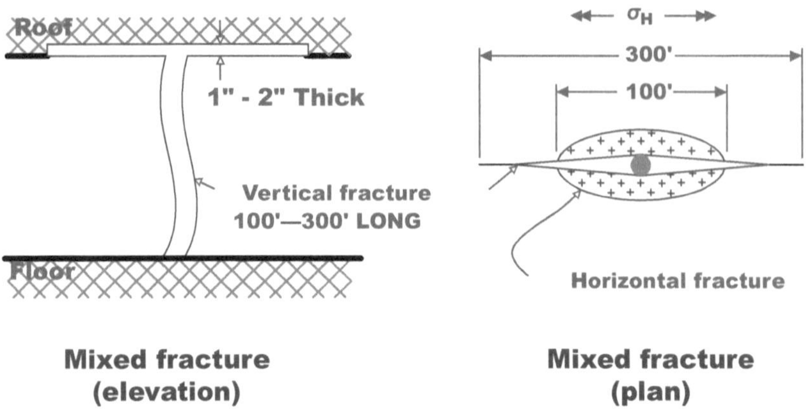

Figure 8.15 A typical mixed fracture.

Such wells are poor producers.

Lineaments and fault zones in coal were normally unable to change the fracture direction. The fine sand (80–100 mesh) effectively plugged them before they could damage the fracking operation.

The best-producing wells had a FG of 1.3–1.6. Wells with a FG of greater than 1.6 tended to have a mixed fracture and were poor producers.

8.8.5 In-mine Observation of a Real-time Fracking Job

It is a rare but valuable opportunity to see in real time how the fracture progresses in a coal seam. A team of observers were stationed in the mine under a supported roof while a well was being fracked. The direction of the fracture was known. After the fracture wing extended to the mine entry (about 500 ft away) the following observations were made:

1. At first you hear the sound of fracking. These are micro-seismic sounds.
2. Next, a tiny frack (less than 1/16 of an inch) went through the roof.
3. Next, the rib of coal in the open entry began to produce some water. The vertical fracture was visible.
4. Next, the fine (80–100 mesh) sand was visible in the fracture.
5. At this point, the frac job ended. The sand was wet but did not have enough water or pressure to extend the fracture.
6. Whenever a fracture ran into an open entry or a previously created fracture, the frac extension of that wing stopped, but the other wing kept growing until the job was completed.

REFERENCES

[1] Nolte KG. Determination of fracture parameters from fracturing pressure decline; SPE Paper 8341, presented at the 54th Annual Meeting of SPE, Las Vegas, Nevada; 1979, p. 23–26.

[2] Perkins TK, Kern LR. Widths of hydraulic fractures. J Petrol Technol 1961;937–49.

[3] Geertsma J, DeKlerk F. A rapid method of predicting width and extent of Hydraulic Induced Fracture. J Petrol Technol 1969;1571–81.

[4] Warpinski NR. Propagating hydraulic fracture; Paper 11648, presented at the SPE/DOE Symposium on Low Permeability Reservoirs, Denver, Colorado; March 14–16, 1983, p. 23–26.

[5] Craft BC, Hawkins MF. Applied petroleum reservoir engineering. Prentice Hall Inc. 1959. p. 437.

CHAPTER 9

Horizontal Drilling in Coal Seams

Contents

Abstract

Vertical wells with hydrofracking reach a production limit at 3000–3500 ft depth for loss of adequate permeability. The deeper coal seams where most of the coalbed methane reserve resides are only amenable to gas production by horizontal drilling from the surface and massive hydrofracking. Horizontal drilling had its beginning about 40 years ago for in-mine horizontal drilling for coal degasification. Because of the space limitations, the drill rig was small and consisted of five components: (1) the drill rig, (2) the auxiliary unit containing the power pack, (3) a drill cutting separation system so water could be reused, (4) a guidance system to guide the drill bit up, down, left or right, and (5) a downhole drill monitoring system that measured the pitch, roll, azimuth, and the distance of the drill bit from the roof of the coal seam. The data was digitized and sent to the surface by a hard wire or as an electromagnetic or acoustic signal. Each of these components is briefly described. Horizontal drilling from the surface requires a much bigger version of the small, permissible, in-mine drill rig. Typical commercial drill rigs and their range of operation are discussed. Drilling procedures both in the mine and on the surface are described. Steel casing schedules for different depth ranges are described. Water and sand schedules for (hydrofracking) a typical 5000 ft long lateral are also described. Assuming two laterals (each 5000 ft long) are drilled into a thick coal seam from the same location and both laterals are hydrofracked at 1000 ft intervals, a production of 4–6 MMCFD can be easily achieved. The process is expensive but if the price of gas is above $5/MCF, it can be highly profitable.

Advanced Reservoir and Production Engineering for Coal Bed Methane.
DOI: http://dx.doi.org/10.1016/B978-0-12-803095-0.00009-0

Horizontal drilling means drilling a long (3000–5000 ft) borehole in the middle of a coal seam that may be 5–60 ft thick. It can be broadly divided into two categories, in-mine horizontal drilling and horizontal drilling from the surface. The former is mainly used for coal mine degasification and the latter is mainly used for commercial gas production. The equipment used for the two methods of drilling is quite different and hence will be discussed separately. The drilling procedures, however, are similar.

9.1 IN-MINE HORIZONTAL DRILLING

This is by far the cheapest and yet the most effective way of degasifying a coal seam prior to mining. The author [1] developed this technique, which can drill a 3- to 4-inch diameter borehole to a depth of 3000–5000 ft. The drill rig is manufactured in the United States by J. H. Fletcher Company in Huntington, WV. Nearly one hundred drill rigs are in use in all major coal mining countries, including the United States, China, India, Australia, and South Africa. Besides coal mine degasification, horizontal boreholes can be used for water drainage and advance exploration for faults, washouts, and other geological anomalies [2,3].

The equipment used to drill long horizontal boreholes can be divided into four major groups: the drill unit, the auxiliary unit, the bit guidance system, and the downhole drill monitor (DDM).

The drill rig provides the thrust and torque necessary to drill 3- to 4-inch diameter boreholes to a depth of 3000–5000 ft. The auxiliary unit provides the high-pressure water to drive a drill motor and flush the cuttings out. It also holds a gas and drill cutting separation system. The bit guidance system guides the drill bit up, down, left, and right as desired in order to keep the borehole in the coal seam. The DDM measures the pitch, roll, and azimuth of the borehole assembly. In addition, it indicates the approximate thickness of coal between the borehole and the roof or floor of the coal seam by using a gamma ray sensor that measures radiation from the roof or floor. The half-depth of gamma rays in coal is typically 8 inches. In recent years, many other uses of in-mine horizontal boreholes have come into practice, such as in situ gasification of coal, improved auger mining, and oil and gas production from shallow deposits.

9.1.1 The Drill Rig

Fig. 9.1 shows the drill unit. It is mounted on a four-wheel drive chassis driven by Staffa hydraulic motors with chains or torque hubs. The tires

Figure 9.1 The drill unit for in-mine horizontal drilling.

are 15 by 18 inches in size and provide a ground clearance of 12 inches. The prime mover is a 50 hp explosion-proof electric motor which is used only for tramming. Once the unit is trammed to the drill site, electric power is disconnected and hydraulic power from the auxiliary unit is turned on. Four floor jacks are used to level the machine and raise the drill head to the desired level. Two 5-inch telescopic hydraulic props, one on each side, anchor the drill unit to the roof.

The drill unit houses the feed carriage and the drilling console. The feed carriage is mounted more or less centrally, has a feed of 12 ft, and can swing laterally by ±17 degrees. It can also sump forward by 4 ft. The drill head has a through chuck such that drill pipes can be fed from the side or back end. The general specifications of the feed carriage are:

High speed:	Torque	5000 lb-in
	RPM = 850	
Low speed:	Torque	11,000 lb-in
	RPM = 470	
Thrust:	30,000 lbs (40,000 lbs pulling out)	
Maximum feed rate:	10–20 ft/min	
Overall dimensions:	length = 16 ft	
	width = 8 ft	
	height = 4 ft	
Maximum tram speed:	1.2 mph	

Jones and Thakur [4] calculated the thrust needed to drill these long horizontal boreholes:

a. In the nonrotary mode

$$\gamma = -2764 + 8.36x_1 + 46.5x_2 + 4376x_3 \tag{9.1}$$

and

b. In the rotary mode

$$\gamma = -236.5 + 418.4x_3 + 1.73x_4 \tag{9.2}$$

c. Minimum torque needed

$$x_4 = 224.2 + 0.22x_1 + 0.3\gamma \tag{9.3}$$

where:

γ = thrust in pounds
x_1 = length of the borehole in feet
x_2 = pressure differential across the drill motor, psi
x_3 = rate of drilling, ft/min
x_4 = torque in lb–inch

In deriving these equations, variables that did not have a significant influence were dropped. In general, the rotary borehole assembly requires much less thrust than the nonrotary borehole assemblies. The azimuth control is very poor with rotary drilling so it is hardly used anymore for long horizontal boreholes.

9.1.2 The Auxiliary Unit

The chassis for the auxiliary unit is identical to the drill unit but the prime movers are two 50 hp explosion-proof electric motors. It is equipped with a methane detector-activated switch so that power will be cut off at a preset methane concentration in the air. No anchoring props are needed for this unit. The auxiliary unit houses the hydraulic power pack, the water (mud) circulating pump, control boxes for electric motors, a trailing cable spool, and a steel tank which serves for water storage and closed-loop separation of drill cuttings and gas.

Fig. 9.2 shows a view of the auxiliary unit.

Fig. 9.3 shows a cross-sectional view of the separation system. The tank is 10 ft × 3.5 ft × 3 ft in size and has two compartments. The inner compartment has sufficient capacity to hold drill cuttings from a 200 ft long hole of 4-inch diameter. Coal fines have a tendency to froth but this is cured with suitable surfactants. At the end of the drilling shift, the vehicle is trammed to a

Figure 9.2 The auxiliary unit for in-mine horizontal drilling.

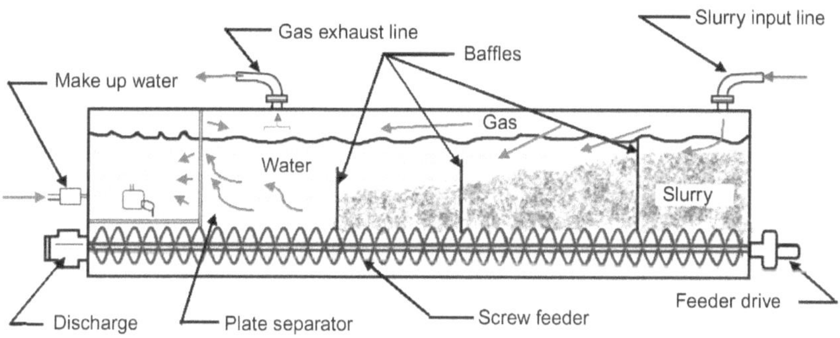

Figure 9.3 The gas and water separation system.

crosscut and the cuttings are discharged by means of a screw feeder. Baffles in the tank collect the large cuttings while fines were initially collected by the plate separator. The latter, however, did not perform entirely satisfactorily and was replaced by a cyclone. Clean water flows to the outer compartment which serves as the storage for fresh water. Float controls in this part of the tank ensure that the correct level of water is always maintained. The low-level float control opens a make-up water valve.

Gas is drawn from the tank via an outlet connected to the underground methane pipeline system. The tank works under slight positive pressure and is designed to withstand a gage pressure of 20 psi.

The water (mud) circulating pump is a triplex, reciprocating pump with a capacity of 70 gpm at 900 psi. In the rotary mode an annulus fluid velocity of 3 ft/s is usually sufficient, but in nonrotary mode the annulus velocity must be increased to 5 ft/s. The pump is driven by a 50 hp electric motor.

The hydraulic power pack consists of a number of hydraulic gear motors capable of delivering 80 gpm of hydraulic fluid at 2500 psi. The working pressure in the system seldom exceeds 2000 psi. Petroleum oil is the recommended fluid for the entire hydraulic system.

9.1.3 The Guidance Systems

When a horizontal hole is started in the middle of a relatively flat, 5–6 ft thick seam the drill bit usually ends up in the roof or floor before reaching 200 ft. In order to drill a deeper hole, it is imperative to guide the bit up and down as needed. In most cases, it is also necessary to guide the bit in the horizontal plane.

To achieve these goals two different modes of drilling, the rotary and the nonrotary modes, were employed. The design of the borehole assembly, i.e., the bit and the first 30 ft of drill column, in either case largely determines the rate of angle build. Figs. 9.4 and 9.5 show the borehole assembly design for the rotary and nonrotary modes of drilling, respectively.

9.1.3.1 Guidance of Rotary Borehole Assembly

In the rotary mode, the drill pipes rotate and all the torque and thrust are provided at the rotary head on the rig. As shown in Fig. 9.4, one stabilizer is used immediately behind the bit and a second is used 10–20 ft behind the first. The first stabilizer also has an internal orienting device for the borehole survey equipment. This stabilizer and 20–30 ft of drill column next to the bit are made of nonmagnetic material so that the borehole survey instruments will not be magnetically affected. Surveying is done with a pumpable tool that measures the pitch, roll, and azimuth of the borehole.

The guidance of the drill bit or, more precisely, the rate of angle built by the bit is actually a factor of two groups of variables: the design of the borehole assembly, and the interaction between the bit and the material

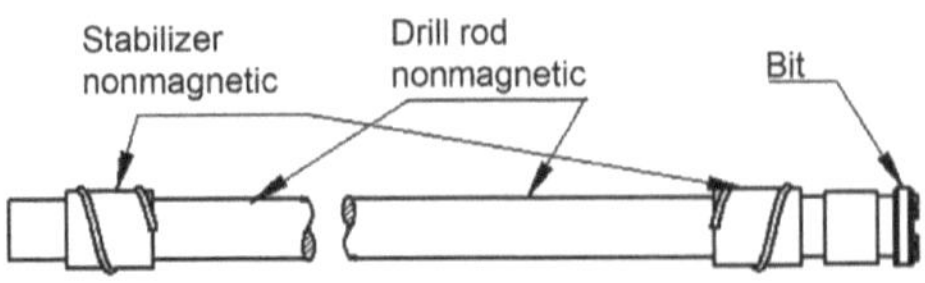

Figure 9.4 Rotary borehole assembly.

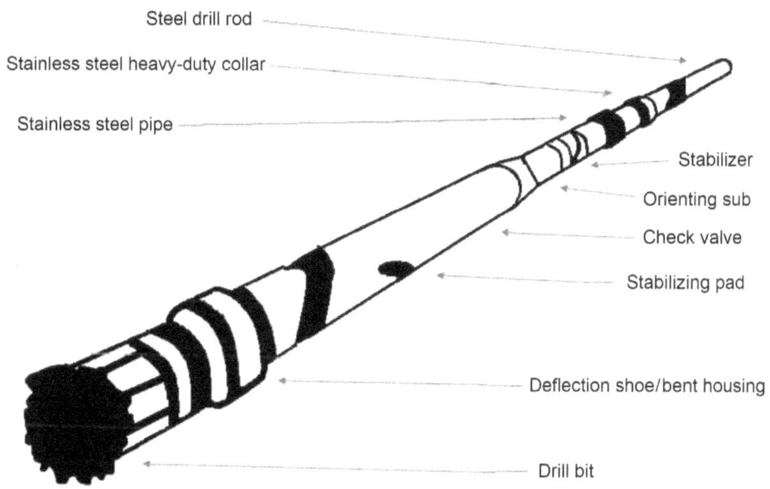

Figure 9.5 Nonrotary borehole assembly.

being drilled. Since coal seams are not uniform, homogeneous strata and bits continuously change their characteristics with wear, it is very difficult to forecast the rate of angle build precisely. For a given type of bit, usually a reasonable rotation speed is selected to yield a penetration rate of 3–5 ft/min and the thrust is varied to make the bit go up and down. At low thrust values the bit pitches down, but at high thrust it will go up. Thrust values and corresponding rates of angle build for a 4 inch diameter drag bit collected for a typical 500 ft of drilling were analyzed using a computer program. A straight-line relationship between the rate of angle build, $\Delta\theta$, and thrust, T, exists as given below:

$$\Delta\theta = 6 \times 10^{-5} T - 0.30121 \quad (9.4)$$

where $\Delta\theta$ is in degrees per 10 ft and T is thrust in lb. In this particular case, the rotary speed was kept steady at 250 rpm and thrust varied from 1000 to 8000 lb. Similar results were obtained in previous studies [5,6].

The three-cone roller and Stratapax bits were also used. They showed similar trends, but the actual rate of angle build varied from bit to bit. With careful selection of drilling parameters, such as the rotary speed and thrust, different kinds of bits can be guided successfully. The drag bit is the easiest to guide but cannot drill through hard rock inclusions in coal. Three-cone roller bits are a little more difficult to guide but will cut through most materials. The life of roller bits is generally less than 1000 ft. Even if the teeth remain sharp, the bearings develop some play and guidance of the bit becomes very poor. Stratapax bits need higher torque but appear to be most suitable for drilling holes deeper than 3000 ft. The biggest drawback of rotary borehole assembly is that it cannot be guided in a horizontal plane. It therefore has a very limited use.

9.1.3.2 Guidance of Nonrotary Borehole Assembly

To overcome the deficiencies of a rotary drilling assembly, a nonrotary assembly was designed. It basically consists of a bit, a deflection device immediately behind the bit, and a downhole motor which runs on the drilling water or mud, as shown in Fig. 9.5. The deflection device was a spring-loaded eccentric sub which exerts a constant force on the side of the bit. The direction of this applied force depends on the orientation of the device and determines whether the bit will be deflected up, down, left, or right. The magnitude of this force and hence the rate of angle build is controlled by the size of the spring. Ideally, the rate of angle build is kept below 0.5 degrees per 10 ft. In coal seams, a side force of 50–100 lb is generally adequate. This device had a tendency to get plugged with coal fines and it was replaced by a "bent housing" of one degree. The drill bit is forced to go up, down, left or right depending on the orientation of the bent housing.

9.1.4 The Downhole Drill Monitor (DDM)

In order to guide the drill bit successfully and contain it in the coal seam, it is essential to know both the position of the bit in relation to the roof and the floor of the coal seam and the pitch of the bit. In the case of nonrotary columns, the roll of the bit and azimuth must also be known so that the deflection device can be properly oriented. Also, the Mine Safety and Health Administration (MSHA)[1] requires that the azimuth of degasification boreholes be plotted on mine maps to prevent inadvertent mining through such holes.

Borehole survey instruments incorporate sensors for the azimuth, pitch, and roll and a coal thickness indicator. The latter indicates the thickness of the coal between the borehole and the floor or the roof, depending on the orientation of the surveying tool. Fig. 9.6 shows the basic components of the survey instrument system, namely the DDM and the readout unit.

The DDM system consists of a downhole survey probe and a portable data collection and display unit situated outside the borehole.

[1] Mine Safety and Health Administration (MSHA) is the US certification agency for all mine equipment.

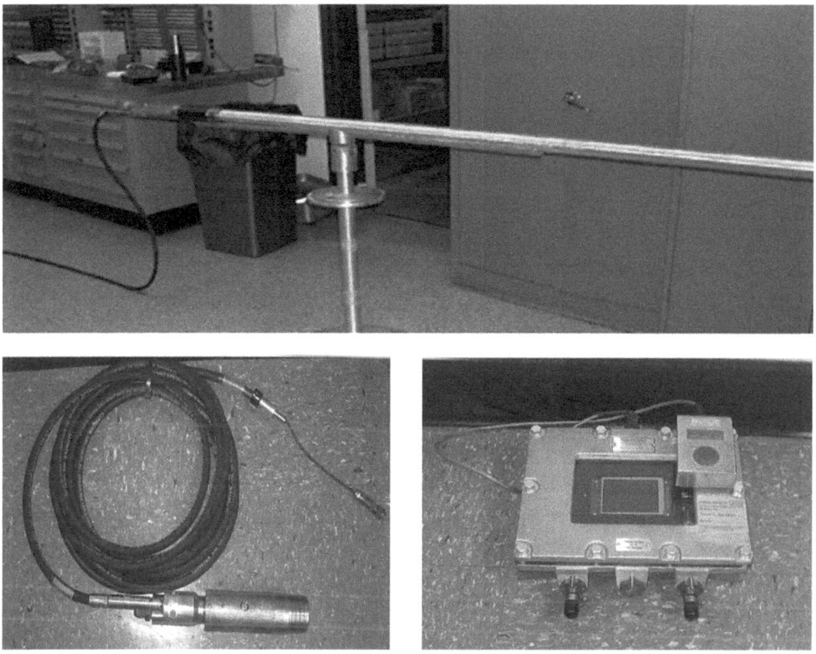

Figure 9.6 The downhole drill monitor.

The downhole survey probe is a battery-powered microprocessor-controlled data acquisition system contained in a 12 ft. long copper-beryllium tube. It is located just behind the downhole motor. The DDM remains downhole until the target depth is reached or until a battery change is needed. A triaxial magnetometer is used to measure the magnetic azimuth. Three accelerometers are used to measure pitch and roll of the drill bit. A solid state gamma detector is used to monitor small amounts of natural gamma radiation emitted from the overlying and underlying shale deposits.

An approximation of roof and floor coal thickness can be made from the observed gamma ray count and the known half-depth value for gamma rays in coal. A built-in computer program controls collection and transmission of data to the collection and display unit. The collected data are digitized and transmitted acoustically through the drill string. Outside the borehole, a magnetic pickup located on the borehole wellhead (or the drill string) receives the signal and displays data sequentially on the display unit. This system has a depth limit of 3000 ft. Recently, a hard-wired communication system was put into use. All drill rods have an insert.

When put together, it provides a solid conductor to transmit data, with a range of well over 5000 ft. It works so well that it has totally replaced acoustic transmission.

All downhole electronic parts are housed in approved explosion-proof aluminum tubing that is watertight and rugged enough to withstand the rigorous downhole environment. The DDM can read pitch, roll, and azimuth with a resolution of 0.1, 1, and 1 degrees respectively. The ranges for pitch, roll, and azimuth are 0–90, 0–360, and 0–360 degrees respectively.

The portable data collection and display unit is a battery-powered, intrinsically safe, MSHA-approved unit for use in return airways of underground mines. The display unit functions as a real-time analyzer to serve the operator in deciding how to orient the bent housing for subsequent drilling and to store various parameters of the borehole being drilled. The storage section of the display unit consists of solid-state memory components with the capability of retaining borehole data which can be taken to the surface and transferred to a larger and more powerful computer. This data can then be used to plot horizontal and vertical profiles of the boreholes. The horizontal profile (plan view) is plotted on mine maps for later use during mine development. The display unit can also be used by the operator to check vertical deviation, horizontal deviation, and drilling parameters such as water pressure and rotary speed if the drilling is done in the rotary mode (i.e., the drilling string is rotated from outside). Data are received by the display unit via a magnetically coupled piezoelectric crystal attached to the wellhead which converts small acoustic signals into electrical signals that are stored in the display unit memory or a hard disc. Each data set received includes the pitch, roll, azimuth and gamma ray counts per minute. After the operator enters a value corresponding to the depth of the borehole, other parameters can be calculated, such as vertical deviation and horizontal deviation with respect to the wellhead. The internal memory of the display unit can store up to 200 sets of borehole data. Any particular data set can be recalled for the operator's review. The CONOCO-developed DDM was licensed in both United States and Australia for commercial production. Since the patent expired, many commercial versions are now available.

9.1.5 Drilling Procedure

For degasification of advancing headings, a drilling site is selected in the outermost headings, which are usually return airways. The drill unit is trammed to the site and set up. The feed carriage is swung laterally until the projected borehole is at 15–20 degrees to the entry headings. The drill head height is adjusted to start drilling in the middle of the coal seam and anchor props are raised to lock the drill unit in position. A surface hole, usually 5 inches in diameter and 20 ft deep, is drilled and a 4-inch O.D. standpipe is grouted using quick-setting cement. A 4-inch gate valve and a commercially made well head are mounted on the standpipe as shown in Fig. 9.7.

This permits safe transport of gas, drill cuttings, and return water to the auxiliary unit through a side outlet without any leakage. A butterfly valve is installed on this line, so that in the event of a sudden influx of gas or ground water, the emission can be contained until arrangements have been made for its disposal. The auxiliary unit can be set up immediately behind the drill unit in the same entry or in the next entry, depending on operating convenience. If gas emission is so high that general body methane concentration cannot be kept below statutory limits, the auxiliary unit can be set up in fresh air.

Drilling is started with a non-rotating borehole assembly as shown in Fig. 9.5 with the bit deflected in the horizontal plane by 5–10 degrees per 100 ft until the azimuth of the borehole is parallel to the entry heading. This is essential because a horizontal hole that deviates very far from the projected headings will provide less effective methane control.

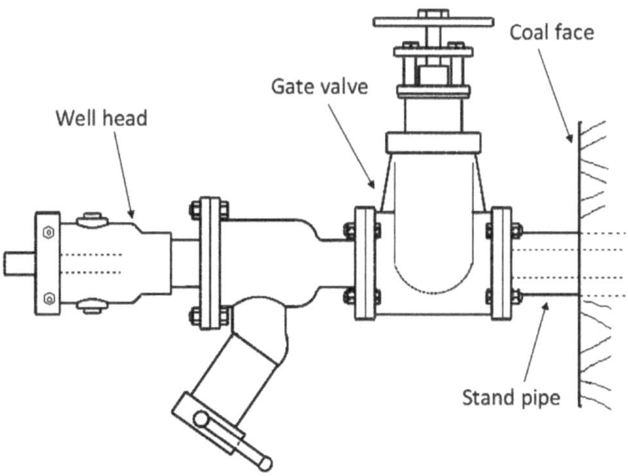

Figure 9.7 The well head assembly.

A borehole survey is taken every 30 ft of drilling. This frequency is recommended for accurate plotting of boreholes, even though the borehole can be maintained in the coal seam with less frequent surveys.

Horizontal holes drilled for advance degasification have ranged from 3 to 6 inches in diameter, but the optimum appears to be between 3 and 4 inches. A bit change is seldom necessary for a 2000 ft hole if Stratapax bits are used. The required length of the borehole is drilled by adding successive drill rods. A check valve is built into the stabilizer immediately behind the bit to prevent water loss, gas emission, and discomfort when the rods are disconnected. On completion of the hole, the rods are withdrawn until the bit is located within the wellhead. The gate valve on the standpipe is closed and the wellhead, complete with the bit, is removed. The standpipe is next connected by means of stainless steel flexible hoses to the underground gas pipeline via a water-gas separator. Flexible hoses are used to accommodate any subsequent ground movements that may occur. The gate valve on the standpipe is now opened and gas is vented to the surface via a vent hole. The gage pressure at the gate valve when closed is usually 4–5 psi. The machine is then trammed out to the next site.

9.1.6 Performance Data

Typical performance of the mobile horizontal drill in the Pittsburgh seam of northern West Virginia is:

Setting up of machine (including water, electric, and hydraulic hook-up)	1 shift
Drilling of anchor pipe hole, cementing and testing	1 shift
Drilling a 2000 ft deep hole and disposing of cuttings	5 shifts
Hook-up of borehole to underground pipeline and tramming out	1 shift
Total time	8 shifts

Only two persons are needed to operate the drill. Typically, one of them is an experienced driller and the other is a helper.

Pitch data are used to plot the vertical profile of the borehole and gamma radiation data off the roof and floor are utilized to project them using the half-depth for the local coal. A typical plot is shown in Fig. 9.8.

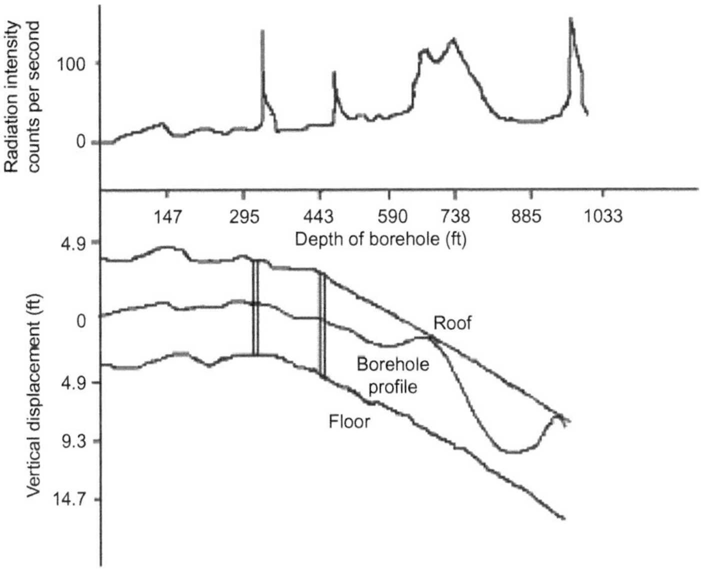

Figure 9.8 Typical plot of a borehole vertical profile.

9.2 HORIZONTAL DRILLING FROM THE SURFACE

The technology for horizontal wells drilled from the surface has been developed in the past 15 years. It is an improvement on the in-mine horizontal drilling procedure. It is mainly used for commercial gas production from shallow or deep coal seams. It is much more expensive than in-mine horizontal drilling. Wells in shallow coal do not need hydrofracking because the natural permeability of the coal is high. In deep coal seams or shale the horizontal laterals are hydrofracked every 250–1000 ft to enhance gas production.

A typical drill site is 4–5 acres in area and has a drill platform for the drill rig, a compressor station for compressed air to get the cuttings out, and a large pond to dump drill cuttings and recover water for reuse. A temporary office is created on-site to provide communications, food, and other facilities for the workers on site. A typical drilling and hydrofracking procedure for a coal seam 8000 ft deep is described here. The coal seam is 60 ft thick and has a gas content of 500 ft^3/ton. It is advisable to complete drilling to the target without any interruptions.

9.2.1 Drilling Procedure

A smaller drill rig, such as a Speedstar 185 with top drive and a hook load capacity of 185,000 lb, is moved to the site and properly anchored.

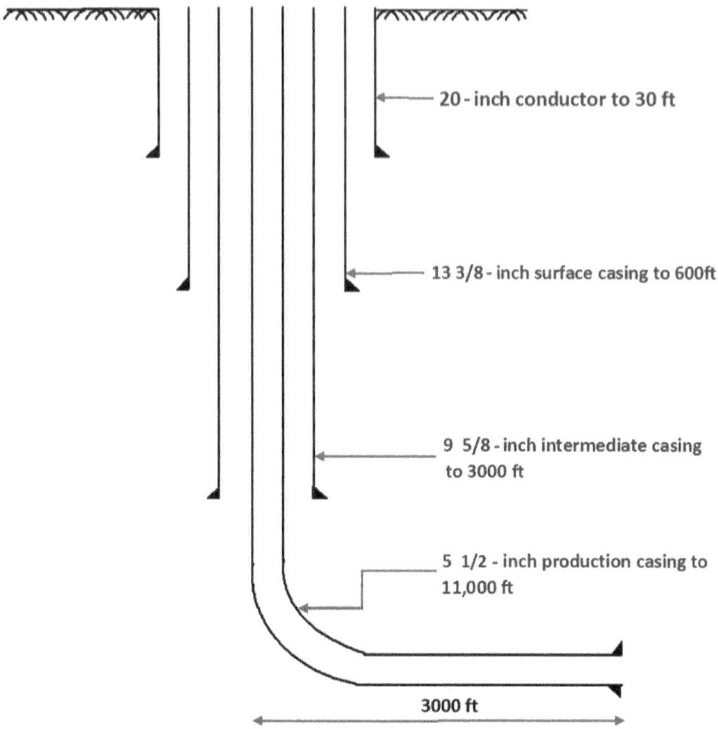

Figure 9.9 A typical horizontal well drilled from the surface.

Fig. 9.9 shows a typical well bore schematic.

First, a 20-inch diameter surface casing is set in place to a depth of 30–40 ft. Then, a 17 ½-inch diameter well is drilled to a depth of about 600 ft (below all known aquifers) and a 13 ⅜-inch casing is set with class A cement. Next, a 12 ¼-inch borehole is drilled to a depth of 3000 ft and the borehole is logged for any minable coal seams. A 9 ⅝-inch casing is set in the well.

Next, the Speedstar rig is moved away from the site and a heavier rig, such as an IDECO Model H-44 double, capable of handling 318,000 lbs hook load is moved to the site. It also has a top drive. A 5000 lb, 9 ⅝-inch casing head is mounted and a blowout preventer is installed.

Next, drilling starts with an 8 ¼-inch polycrystalline diamond (PCD) bit with 6 ½-inch drill collars (rods). The well and flow lines are pressure tested and all safety protocols are completed. The production well is drilled to a target depth well below the target coal seam (usually 100–200 ft below). The well is logged again to choose the location where the deviated borehole will start. Assuming that the kickoff point will be at 7500 ft, the bottom of the vertical well is cemented to a depth of 6000 ft. The directional borehole assembly is lowered in the well and the well is drilled to the target kickoff point of 7500 ft. Next, the curve is drilled with foam mist to make the vertical well a horizontal one. The drill cuttings indicate if the well has entered the coal seam. The rate of angle build is 8–12 degrees per 100 ft. A 2 degree bent housing is used for this purpose. Directional control of the well is usually provided by professional directional drillers. The horizontal lateral is drilled with a PCD bit of 8 ⅝-inch diameter and a mud-driven motor (a Moyno pump in reverse).

In coal seams, all drilling is done with foam but in shale they use a 12–14 ppg mud (12–14 lbs of mud in a gallon of water). The horizontal drilling continues until the target depth is reached. For a 3000 ft lateral, the target depth would be about 11,000 ft. The drill string is tripped out and a 5 ½-inch, 20 lb/ft, P-110 casing is cemented in the entire well.

9.2.2 Hydrofracking of the Lateral

The approximately 3000 ft long horizontal lateral is next hydrofracked through perforations in five sections to enhance the permeability. As discussed earlier in the book, slick water (fresh water with a friction reducer, such as, polyacrylamide) is used. The hydrofrac should be properly designed using the theories discussed in this book. Data for a typical well in Devonian shale is presented in Table 9.1. No such hydrofracking has been done in a coal seam so far but the process would be very similar.

This is a massive hydrofracking job using over 4 million gallons of water and 3.35 million pounds of sand.

In a coal seam, the laterals should be drilled parallel to σ_h (the minor horizontal stress) such that the fractures will be parallel to σ_H or the face

Table 9.1 Hydrofracking a 3000 ft lateral in five stages

Stage	Fluid volume (bbl)	Sand (lbs) 100 mesh	Sand (lbs) 40/70 mesh	Rate bbl/min
1	20,000	180,000	500,000	102
2	19,000	170,000	510,000	105
3	21,000	190,000	480,000	101
4	18,000	180,000	470,000	106
5	19,000	160,000	510,000	106
Total	97,000	880,000	2,470,000	

cleat, yielding higher gas productions. Assuming two laterals (each 5000 ft long) are drilled from the same location and both laterals are hydrofracked as discussed above, a total of 20 fractures, each of 2000 ft length, are created. The total length of 40,000 ft can produce 4–6 MMCFD assuming a specific gas production of 10–15 MCFD/100 ft. The specific gas production is a characteristic of the coal seam and the completion procedure. It will be discussed in Chapter 10, Coalbed Methane Production From Shallow Coal Reservoirs.

REFERENCES

[1] Thakur PC, Poundstone WN. Horizontal drilling technology for advance degasification. Min Eng 1980;676–80.
[2] Thakur PC, Davis JG. How to plan for methane control in underground coal mines. Min Eng 1977;41–5.
[3] Thakur PC, Dahl HD. Horizontal drilling—a tool for improved productivity. Min Eng 1982;301–4.
[4] Jones E, Thakur PC. Design of a mobile horizontal drill rig, Proceedings of the 2nd Annual Methane Recovery from Coalbeds Symposium, Pittsburgh, PA; 1979, p. 185–193.
[5] Cervik, et al. Rotary drilling holes in coalbeds for degasification. RI 8097: US Bureau of Mines; 1975.
[6] Rommel R, Rives L. USBM Contract No. H0111355 Advanced techniques for drilling 1,000-ft small diameter horizontal holes in coal seams. Tulsa, Oklahoma: Fenix and Scisson Inc.; 1973.

CHAPTER 10

Coalbed Methane Production From Shallow Coal Reservoirs

Contents

Abstract

Coal seam gas reservoirs are very different from conventional oil and gas reservoirs. Most coalbed methane (CBM) reservoir properties are depth-dependent. Hence, all CBM reservoirs are classified as either shallow, medium-depth, or deep reservoirs. Four US coal basins are classified as shallow reservoirs, namely the Powder River Basin, Cherokee Basin, Illinois Basin, and Northern Appalachian Basin. Together they produce 22% of the current US CBM production.

The Powder River Basin is a shallow (less than 1000 ft deep) reservoir made up of very thick coal seams with about 75 ft^3/t gas content. A typical CBM well is drilled into the coal seam and hydro-jetted prior to production. Only rarely is the well stimulated. Average production per well is 150 MCFD. The basin is producing 280 BCF/year. The basin lies mostly in Wyoming and Montana.

The Cherokee Basin is a small basin located at the junction of Oklahoma, Kansas, and Missouri. The coal seams have a gas content of 200 ft^3/t. Vertical wells with hydraulic stimulation produce 250 MCFD/well. The field is not well developed and it produces only 5 BCF/year.

The Illinois Basin lies in the states of Illinois, Indiana, and Kentucky and is the least developed CBM basin. Vertical drilling with hydrofracking has been tried but the gas production is disappointingly low. Annual gas production is less than 1 BCF/year.

The Northern Appalachia Basin lies mostly in Pennsylvania and West Virginia and has a gas reserve of 61 TCF. Seams to a depth of 1200 ft are produced by horizontal drilling from surface and vertical drilling and hydrofracking. Horizontal drilling typically produces 300–600 MCFD while vertical wells produce 0–75 MCFD.

For commercial gas production, horizontal drilling from the surface with laterals of 3000–5000 ft is the only way to go. Hydrofracking is generally not needed because it is not effective at shallow depths.

Advanced Reservoir and Production Engineering for Coal Bed Methane.
DOI: http://dx.doi.org/10.1016/B978-0-12-803095-0.00010-7

Global Coalbed methane (CBM) resources as shown in Chapter 1, Global Reserves of Coal Bed Methane and Prominent Coal Basins, Table 1.3, are huge. Production strategy varies from basin to basin, and even within the same basin depending on local reservoir properties. Since all coal seam reservoir properties are depth-dependent, CBM resources can be classified into three broad categories to identify the best production technique for different ranges of depth: shallow, medium-depth, and deep reservoirs. The main characteristics of these reservoirs are summarized in Table 10.1.

The United States, with its vast CBM reserves at about 400 TCF and a peak production of 2 TCF/year, uses a variety of production techniques that can be emulated in the rest of the world. Owing to serious competition from cheap shale gas, production has declined to 1.4 TCF/year in 2015 and is distributed as shown in Table 10.2 [1]. Fig. 10.1 shows the current US production fields.

Table 10.1 Reservoir characteristics

Reservoir type	Approximate depth (ft)	Permeability (md)	Stress field
Shallow	500–1500	10–100	$\sigma H > \sigma h > \sigma \nu$
Medium-depth	1500–3300	1–10	$\sigma H > \sigma \nu > \sigma h >$
Deep	>3300	0.1–1.0	$\sigma H > \sigma \nu > \sigma h >$ [a]

[a]There are some exceptions to this general rule as discussed in Chapter 5.

Table 10.2 US Coalbed methane production distribution (2015)

Reservoir type	Basin	Production (BCF/year)
Shallow	(a) Powder River Basin	280
	(b) Cherokee Basin	5
	(c) Illinois Basin	1
	(d) Northern Appalachian Basin	10
Medium-depth	(a) Central Appalachian Basin	94
	(b) Warrior Appalachian Basin	52
	(c) Raton Basin	105
	(d) Arkoma Basin	100
Deep	(a) San Juan Basin	650
	(b) Uinta Basin	40
	(c) Piceance Basin	5
	(d) Green River Basin	20
Total production		1.362 TCFD

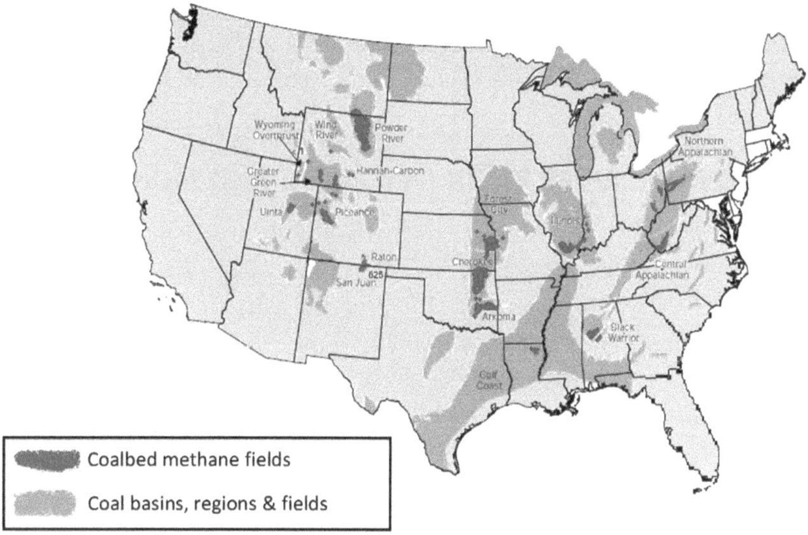

Figure 10.1 US CBM production fields.

Table 10.3 Specific gas emissions for coal seams

Coal seam	Depth (ft)	Rank	Specific gas production MCFD/100 ft
Pittsburgh	500–1000	High vol. bituminous	15.00
Pocahontas No. 3	1400–2000	Low vol. bituminous	8.00
Blue Creek/ Mary Lee	1400–2000	Low vol. bituminous	9.00
Pocahontas No. 4	800–1200	Medium vol. bituminous	5.00
Sunnyside	1400–2000	High vol. bituminous	9.00

10.1 SPECIFIC GAS PRODUCTION FOR A COAL SEAM

Thakur [2] defined the term "specific gas production" for a coal seam to estimate gas production. Specific gas production for a coal seam is a characteristic of the coal seam and is measured by the initial gas production from a 100-ft horizontal borehole drilled in the coal seam. It is a useful parameter to estimate production from either a vertical well (assuming the fracture length is known) or a horizontal borehole drilled from mine workings or the surface. Table 10.3 provides the data for some well-known coal seams of the United States [3].

These numbers will be used to estimate the initial production from various basins. The production techniques used and resultant gas production rates will be discussed for each of the 12 basins listed in Table 10.2.

10.2 POWDER RIVER BASIN (WYOMING AND MONTANA)

The geology and gas reserves were discussed in Chapter 1, Global Reserves of Coal Bed Methane and Prominent Coal Basins. A map of the basin is shown in Fig. 10.2. Most of the current gas production is obtained from wells that are 200–600 ft deep. The coal is highly permeable and has a reservoir pressure gradient of 0.26–0.29 psi/ft, nearly 70% of the hydrostatic gradient of 0.434 psi/ft. Net coal thickness varies from 170–300 ft, but the gas content is low at 70–80 ft^3/t. The coal is of low rank, ranging from lignite to sub-bituminous coal.

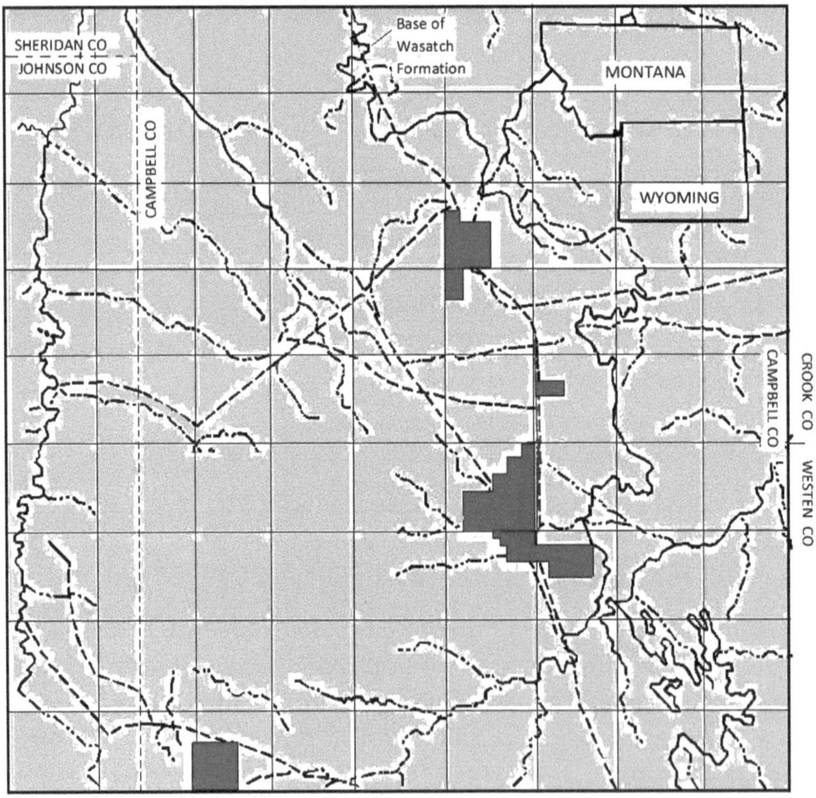

Figure 10.2 Major CBM fields of the powder river basin.

Most CBM wells are drilled and completed in a single day using a conventional truck-mounted water-well drilling rig. Fort Union coal seams are generally completed open hole and unstimulated. In some cases, a small volume hydrofracking (less than 15,000 gallons of water and about 10,000 pounds of sand) has been done with modest improvement in production. This is compatible with the theoretical conclusion that hydrofracking of shallow coal will create a horizontal fracture with only marginal increase in gas flow rates. Gas production from such wells ranges from 50–300 MCFD. The coal seams are highly saturated with water. Typical water production is 200–400 bpd, but the water is of good quality (<5000 ppm of TDS) and can be beneficially used.

An alternative technique to produce a much larger quantity of gas from a single location would be to sink a shaft of 15–16 ft in diameter and drill horizontal boreholes at the bottom of the shaft. Assuming six laterals of 4000 ft are drilled as shown in Fig. 10.3, a production of 1 MMCFD can be obtained. A specific gas production of 4 MCFD/100 ft is assumed. The gas production rate will improve as water drains out and then decline, and production may last for over 20 years. The choice of production technique will depend on the economics of the option. The basin is currently producing 280 BCF/year from over 10,000 wells.

10.3 THE CHEROKEE BASIN

This is a smaller basin located near the Oklahoma-Kansas-Missouri State boundary (see Fig. 10.1). The depth of the productive coal seams is 600–1200 ft with an average gas content of about 200 ft^3/t.

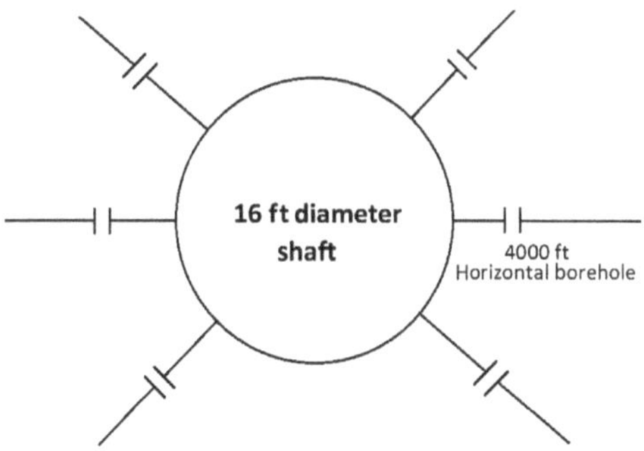

Figure 10.3 Horizontal production boreholes at shaft bottom.

Vertical wells with hydraulic stimulation have been used to produce up to 250 MCFD/well. Gas production can be significantly improved by drilling long (4000–5000 ft) horizontal laterals from the surface. The coal seams (Weir-Pittsburg) have good permeability. The specific gas production is estimated at 15 MCFD/100 ft. Assuming four laterals of 5000 ft are drilled from a common production well, an initial production of 3 MMCFD can be realized. Details of the drilling procedure will be the same as in the Illinois Basin, to be discussed next. At present, the total gas production is estimated at 5 BCF/year.

10.4 THE ILLINOIS BASIN

Fig. 10.4 shows the coal reserve and potential gas production areas of this basin. The gas content of the coal varies from 30–150 ft^3/t with a total gas reserve of about 21 TCF to a depth of 1500 ft.

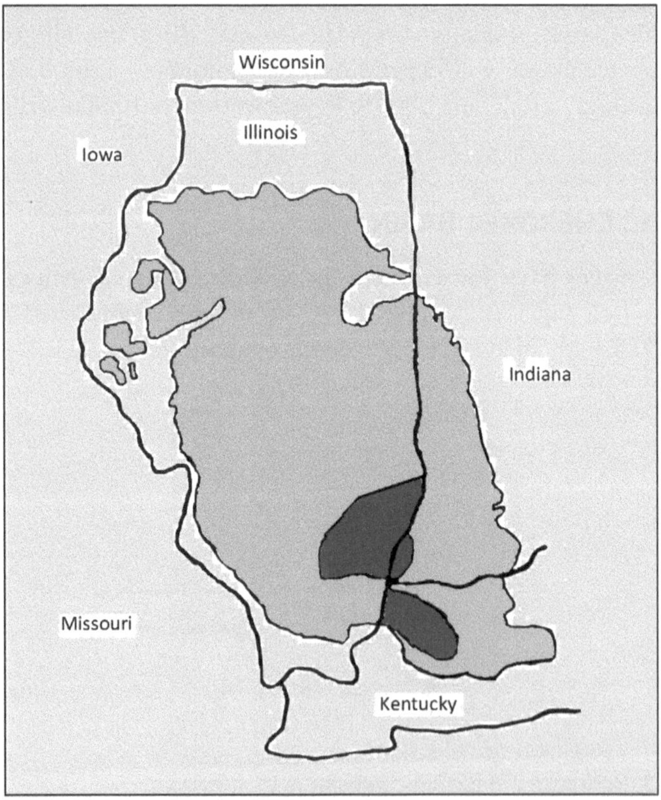

Figure 10.4 Illinois Basin CBM production fields.

The target coal seams are Herrin #6 and Herrin #5 (Springfield), which are about 6 ft thick. The coal rank is mostly hvC to hvA bituminous. Vertical drilling and hydrofracking is the main production technique used so far. As discussed, this is a very ineffective way to produce gas because the fractures are horizontal. In most cases, the coal gets separated from the roof and the sand is deposited there.

The best technique to produce gas from this basin would be to drill a production well and case it with 9 5/8-inch casing. The coal seam is reamed to a larger diameter of 4–6 ft by a high-pressure water jet. Next, horizontal boreholes are drilled into it from a distance of 4000–5000 ft as shown in Fig. 10.5. Assuming a total of 20,000 ft is drilled and the coal seam has a specific gas production of 6 MCFD/100 ft, an initial production of 1.2 MMCFD can be realized. The coal seams are highly saturated with water and, therefore, a high water production rate is anticipated. A submersible electric pump is recommended for water removal from the production well. The well spacing is one well/square mile. The gas reserve for each site is about 600 MMCF. The current gas production from the basin is about 1 BCF/year, but it can be substantially increased.

10.5 NORTHERN APPALACHIAN BASIN

This basin contains 61 TCF gas in 352 billion tons of coal. Fig. 10.6 shows the geographical location of the basin. The total thickness of coal

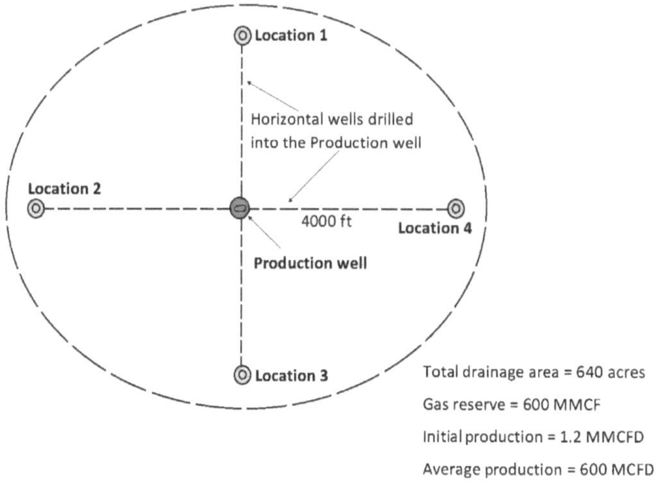

Figure 10.5 Gas production scheme in the Illinois Basin.

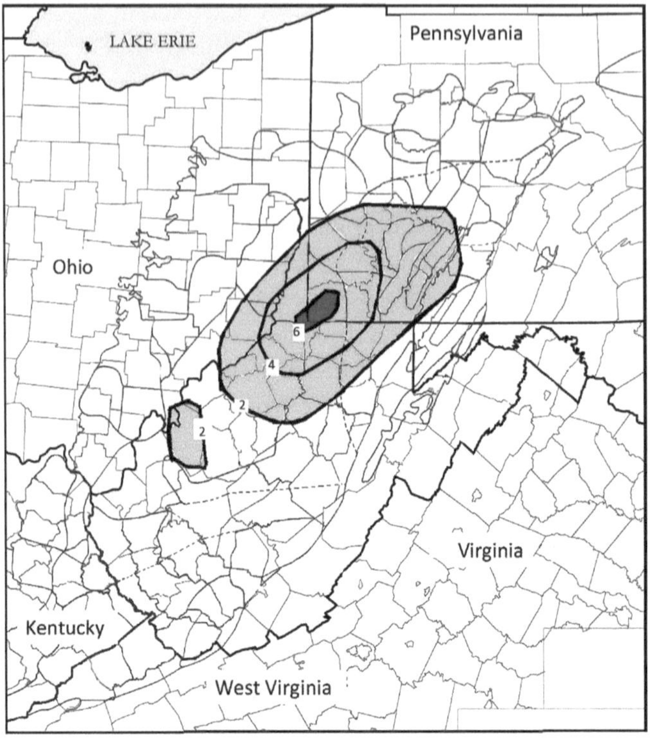

Figure 10.6 The Northern Appalachian Coal Basin.

Table 10.4 Major coal seams of Northern Appalachian Basins

Seam	Rank	Depth (ft)	Total gas reserve (TCF)
Pittsburgh	High vol. A bituminous	500–1200	7.0
Freeport	High vol. A bituminous	400–1600	15.5
Kittanning	High vol. A bituminous	800–1600	24.0
Brookville/Clarion	High vol. A bituminous	800–2000	11.0

to a depth of 2000 ft is 28 ft, but individual coal seam thickness ranges from 5 to 15 ft.

Major coal seams and their gas content are shown in Table 10.4.

The Kittanning and Brookville seams can be treated as medium-depth coal seams, and gas production techniques for them will be discussed in the Chapter 11, Coal Bed Methane Production from Medium-depth Coal Reservoirs.

The Freeport and Pittsburgh seams fall under the shallow coal category. There are three different production techniques currently in use.

1. Vertical drilling and hydrofracking. In Indiana County, Pennsylvania, many wells are completed at a depth of 800–1000 ft. They produce an average of 75 MCFD with 300 bpd of water. The quality of the water is good. The water produced is collected in treatment ponds and released into the local streams after the solids have settled down in the ponds.
2. Horizontal boreholes drilled from the surface. Another operator used a pattern of horizontal boreholes drilled from the surface in Green and Washington Counties, Pennsylvania, and Monongalia County, West Virginia to produce from the Pittsburgh coal seam. Fig. 10.7 shows a typical pattern. Assuming that the main lateral was 3000 ft long and two parallel laterals of equal length were drilled about 1000 ft apart, the total length of 9000 ft produced only 300–400 MCFD yielding a specific gas production of 3–5 MCFD/100 ft. The apparent low yield is due to the fact that the three laterals communicated with each other (this coal seam is highly permeable) and behaved like a single 3000-ft borehole, yielding 9–15 MCFD/100 ft of borehole. The gas production is much better than that from vertical wells drilled into the coal seam and hydrofracked for gas production. The gas production from hydrofracked wells was poor at 0–50 MCFD because all fractures were horizontal.

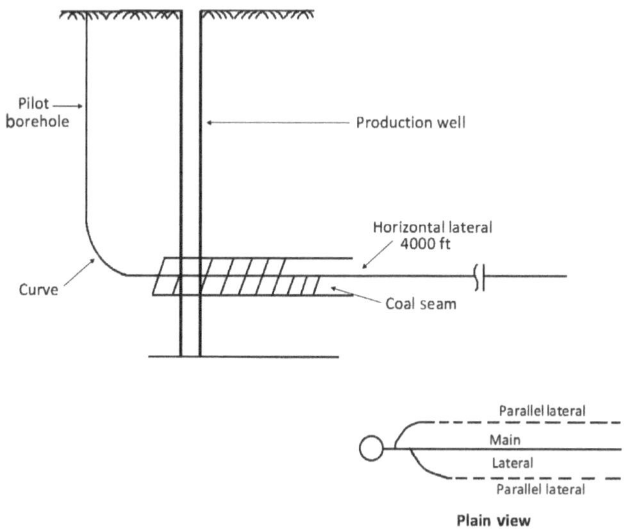

Figure 10.7 Gas production scheme in Southwest Pennsylvania.

3. The best production technique for the Pittsburgh and Freeport coal seams is the technique already described for the Illinois Basin. A production well should be drilled with 1000-acre spacing, and 5–6 horizontal boreholes should be drilled into it from a distance of 4000 ft. The directional guidance system is so good that the boreholes can drill into the production well from this distance. Good directional drillers can maintain the entire borehole in the coal seam. Assuming a total of 24,000 ft of horizontal borehole is drilled into the production well, an initial production of 3.6 MMCFD can be realized. Water production from these coal seams is not high (10–20 bbd). A rod-and-beam pump with a timer can be used to pump the water out. Since these boreholes are drilled in a mineable coal seam, they need to be plugged prior to mining. Each borehole can be independently plugged with class A cement. When all production ceases, the production well is plugged for safety in mining.

For a medium-size gas project producing 50 MMCFD, nearly 30 production wells will be needed. A production area of 50 square miles can sustain this production rate for over 20 years.

REFERENCES

[1] EIA. Coal Bed Methane Proved Reserves. Washington, DC: Energy Information Administration; 2014. (www.eia.DOE.gov).
[2] Thakur PC. Optimized degasification and ventilation for gassy coal mines, The Ninth International Mine Ventilation Congress, New Delhi, India; 2009.
[3] Thakur PC, et al. Horizontal drilling technology for coal seam methane recovery, The Fourth International Mine Ventilation Congress, Brisbane, Australia; 1988.

CHAPTER 11

Coalbed Methane Production From Medium-Depth Coal Reservoirs

Contents

Abstract

There are four basins at depths of 1500–3000 ft that currently produce nearly 25% of all US CBM. They are the Central Appalachian Basin, the Warrior Basin, the Arkoma Basin, and the Raton Basin. Vertical drilling with hydrofracking is exclusively used for gas production. The coal seams are usually thin (5–6 ft) and multiple coal seams are stimulated in a single well.

1. The Central Appalachian Basin: The most productive area is located in southwestern Virginia and southern West Virginia. Massive hydrofracturing of coal started here in 1984. A well completed in multiple coal seams (total thickness ≥ 15 ft) can produce initially at 450 MCFD and yield 1–1.5 BCF gas before abandonment. The basin has a CBM reserve of 21 TCF and the current annual production is 98 BCF.
2. The Warrior Basin: It lies mostly in Alabama and Mississippi, but only the Alabama side is well developed. It has a reserve of 21 TCF and an annual production of 51 BCF. Production peaked at 92 BCF/year in 1992. The basin is extensively drilled and new drilling has declined.
3. The Arkoma Basin: It is located in Oklahoma and Arkansas. The basin is medium-sized at 13,500 square miles, but the coal reserve is low at a mere 8 billion tons. The coal seams are thin but highly gassy. The gas reserve is estimated at 3 TCF. Current annual production is 100 BCF.
4. The Raton Basin: It is located in southeastern Colorado and northwestern New Mexico. The total gas reserve is 11 TCF. Annual gas production is 105 BCF. The coal seams are thick and gassy. There is room to improve the hydrofracking technique for higher gas production per well.

Advanced Reservoir and Production Engineering for Coal Bed Methane.
DOI: http://dx.doi.org/10.1016/B978-0-12-803095-0.00011-9

As shown in Fig. 10.1, there are four coal basins in the United States where coalbed methane (CBM) is being produced from a depth horizon of 1500–3000 ft:

1. the Central Appalachian Basin
2. the Warrior (Southern Appalachian) Basin
3. the Arkoma Basin and
4. the Raton Basin.

Together, they produce nearly 25% of total US CBM production.

11.1 THE CENTRAL APPALACHIAN BASIN

This is a narrow, northeast-trending basin covering an area of 23,000 square miles in Tennessee, Kentucky, Virginia, and West Virginia. Only a 5000 square-mile area in southwestern Virginia and southern West Virginia has a high potential for CBM production (Fig. 11.1). The rank of coal is medium to low volatile bituminous with gas contents of 300–650 ft^3/t. Individual coal seams are 4–6 ft thick, but total thickness of coal to a depth of 2500 ft is nearly 30 ft. Not all coal seams are amenable to gas production because some are too thin (less than 2 ft).

Gas-in-place in this highly productive area exceeds 25–30 MMCF/acre (16–19 BCF/section). The quality of gas is very good with 95% methane

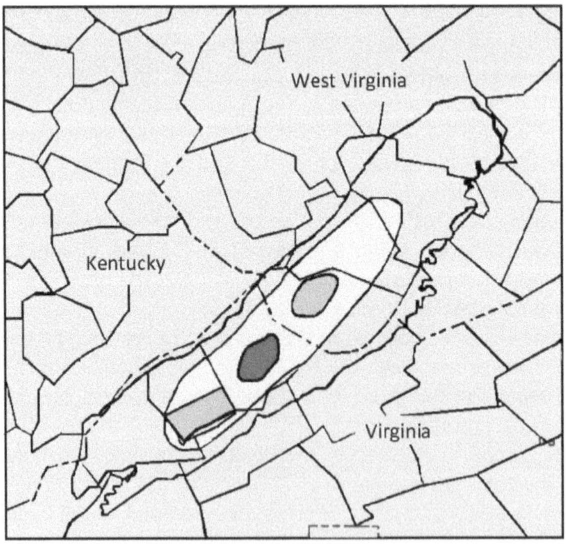

Figure 11.1 The Central Appalachian Basin.

Table 11.1 Characteristics of producible coal seams in Central Appalachian Basin

Coal seam	Depth (ft)	Thickness (ft)	Gas content (ft^3/t)
Jaeger	400–600	5	400–600
Beckley/Firecreek	600–1200	5	500–550
Pocahontas #4	1200–1500	4–6	300–600
Pocahontas #3	1500–2500	5–6	450–650

and only 4% noncombustibles. The characteristics of the producible coal seams are shown in Table 11.1.

Initial production from this field started when Thakur hydrofracked the first well on November 15, 1984, in Buchanan County, Virginia. Only the Pocahontas #3 seam was stimulated, with 100,000 gallons of slick water and 120,000 pounds of sand at 30 bbl/min. Initial production ranged from 100–250 MCFD. Encouraged by the early results, multiple seams were hydrofracked in a single well using 70% nitrogen foam. The production from a typical well is shown in Fig. 11.2.

This well had a peak production of 450 MCFD with about 15% annual decline. It was drilled on a 60-acre spacing and was completed in three horizons of coal with a total thickness of about 20 ft. It was a 70% N_2 foam fracture with 60,000 gallons of water, 1.5 MMCF of nitrogen, and nearly 200,000 pounds of sand that was pumped at 35 bbl/min. Gas production took 2 years to peak and will take 13 years to decline to abandonment with an estimated cumulative production of 1.2 BCF. Total gas-in-place for this well is 1.8 BCF. Hence, the recovery is about 67%, which is very good. Such high recovery is possible because the coal seam has a high diffusivity.

One hundred twenty-five wells like this over an area of 7500 acres can create an initial gas production of 50 MMCFD. A commercial venture in this area is producing 72 BCF/year from a 250,000-acre reservoir. This is a very good example of a successful CBM project. Total CBM production from this basin is 98 BCF/year. The total gas reserve in the basin is 21 TCF. Several operators are engaged in CBM production from this basin.

11.1.1 Horizontal Boreholes Drilled From the Surface

In the shallow parts of the Central Appalachian Basin, horizontal drilling from the surface has been employed for significant gas production. The coal seam is dry and has a depth of less than 1500 ft. The permeability is

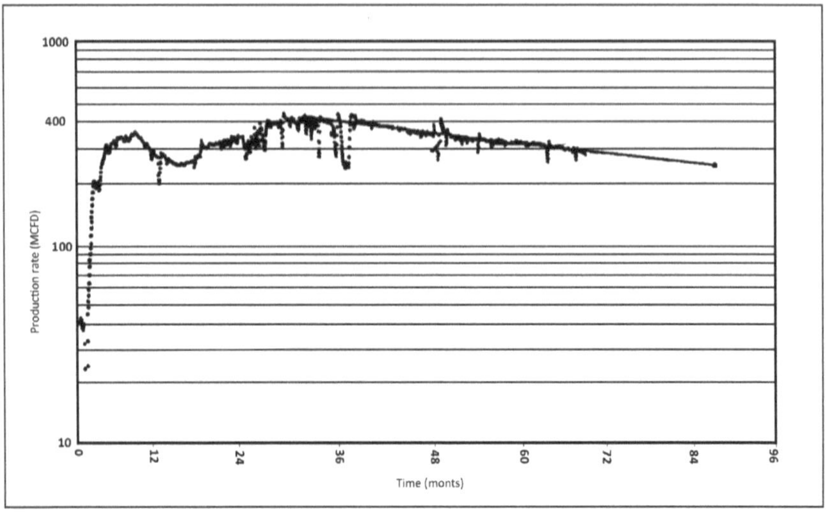

Figure 11.2 Typical production from a multi-seam well in the Central Appalachian Basin.

high, and a specific gas production of 6–8 MCFD/100 ft was achieved in the Raleigh County of West Virginia. As a general rule, horizontal drilling may be feasible to a depth of 2000 ft, but beyond that, the permeability is poor and hydraulic stimulation becomes necessary. Vertical drilling and hydrofracking of multiple coal seams (with a total thickness of 15–30 ft) becomes economically more attractive when the depth exceeds 2000 ft.

11.2 THE WARRIOR BASIN

The Warrior Basin is triangular and covers an area of 12,000 square miles equally divided between Alabama and Mississippi. Only the Alabama side of the basin is well developed. In the mid-1990s, a peak production of 92 BCF/year was obtained from nearly 2000 wells, which has declined to 51 BCF/year in 2015. The coal and gas reserve in the Alabama side of the basin are 62 billion tons and 21 TCF respectively. At one time, as many as 20 companies were engaged in CBM production. Fig. 11.3 shows the geographical layout of the basin.

Major coal seams with depth and gas content are listed in Table 11.2.

The rank of coal ranges from hvA to low vol. bituminous. Permeability declines with depth, as discussed earlier.

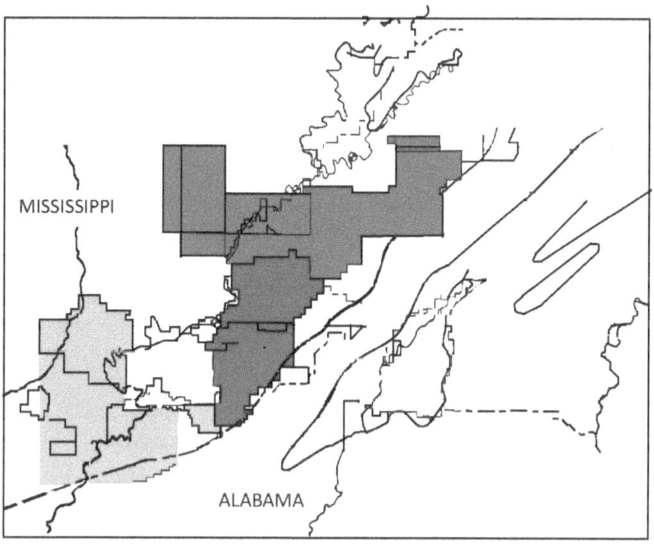

Figure 11.3 CBM-producing areas in the Warrior Basin.

Table 11.2 Main coal seams of the Warrior Basin

Seam	Depth (ft)	Gas content (ft^3/t)	Permeability (md)
Pratt	700–2200	200–400	10–15
Blue Creek/Mary Lee	1200–2800	400–500	5–15
Black Creek	1200–3300	450–550	1–10

Vertical drilling with hydrofracking of multiple coal seams is the most common method of gas production.

Fig. 11.4 shows a typical completion in three production coal seams. The bottom coal seam is typically hydrofracked in the open hole (no casing), but the upper coal seams are hydrofracked through perforations in the casing. Average gas production from a well ranges from 100–150 MCFD. This is less than 50% of the gas production per well in the Central Appalachian Basin. The hydrofracking procedure here needs to be optimized to increase the gas production.

11.2.1 The Mississippi Extension of the Warrior Basin

Exploratory drilling in Clay County shows promising results. Major coal groups are West Point, Sand Creek, Houlka Creek, and Lime Creek, in descending order. There are 17 seams in total with a total coal thickness of 20 ft. The rank of coal is hvA. The coal resource and methane resource

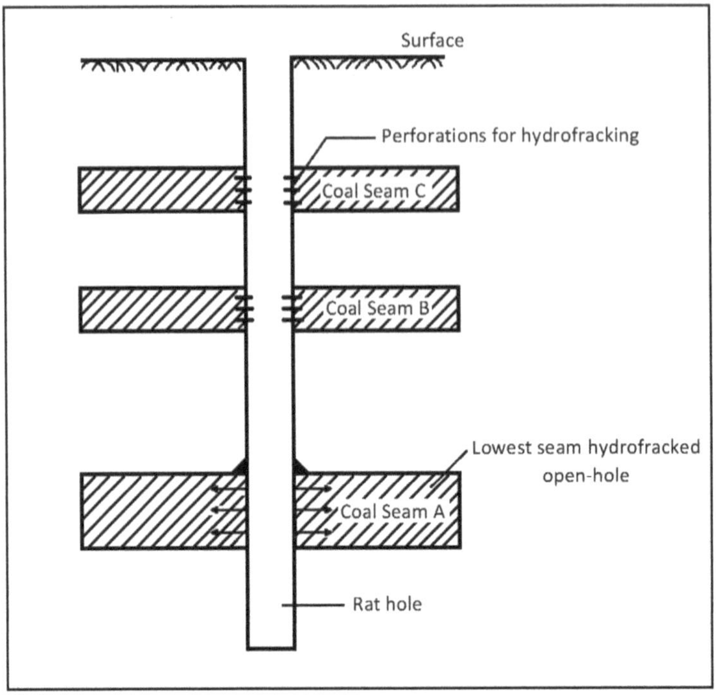

Figure 11.4 A typical multiple completion in three coal seams.

are estimated at 5.2 billion tons and 1 TCF, respectively, to a depth of 3700 ft. The gas content is lower than that of the coal deposits of the Alabama side, at about 200 ft^3/t. Vertical drilling with hydrofracking is recommended for commercial gas production. To optimize gas production, additional reservoir data will be necessary.

11.3 THE ARKOMA BASIN (KANSAS, MISSOURI, ARKANSAS, AND OKLAHOMA)

The Arkoma Basin is the southern part of the larger Western Interior Coal Region, which also includes the Forest City Basin and Cherokee Basin (discussed earlier) to the north.

Fig. 11.5 shows the geographical location of all three Basins. Arkoma Basin of east-central Oklahoma and West-Central Arkansas contains the deepest and gassiest coal seams. It is nearly 13,500 square miles in area and has a coal reserve of about 8 billion tons. Major coal seams in ascending order are (1) Hartshorne, (2) Savanna, and (3) Boggy formations.

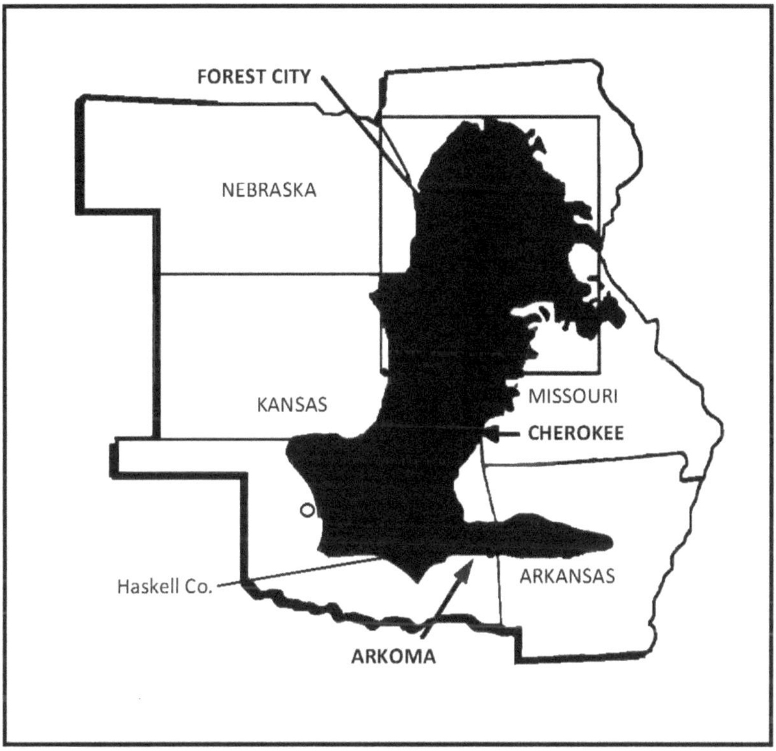

Figure 11.5 Map of the Arkoma, Cherokee, and Forest City Basins in the Western Interior Coal Region.

The rank of coal increases from West (HvA Bituminous) to low vol. coal and anthracite to the East. The gas content of coal seams ranges from 200–670 ft^3/t, the latter number is the highest recorded in the United States. Assuming an average gas content of 400 ft^3/t, the estimated methane reserve is about 3 TCF. The thickness of Hartshorne seam varies from 3–9 ft at a depth of 1500–2000 ft.

The basin has been drilled with vertical wells since 1970 on a limited basis. Several operators began to drill for commercial production in 1989. The current gas production is about 100 BCF/year. Vertical drilling and hydrofracking procedures are similar to those in the Central Appalachian Basin, but the process does not appear to have been optimized. Water, linear gel and nitrogen foam have been used to hydrofrack. Most hydrofracking jobs are small (30,000–60,000 pounds of sand). Gas production is 50 MCFD/well with 10–50 bbl/day of water. The Hartshorne coal seam does not produce much water.

A preferred completion procedure for the basin would be to drill and hydrofracture wells in areas where approximately 10–20 ft of coal is present. The gas content is likely to be 500 ft^3/t at depths of 1800–2500 ft. The lowest coal seam should be completed open-hole, and overlying coal seams should be hydrofracked through perforations. The hydrofracking job should use 150,000–200,000 gallons of slick water with 200,000 pounds of sand. The hydrofracking process should be optimized based on prior results. An average production of about 300 MCFD can be realized under these conditions.

The total gas production from the basin is 100 BCF, but it can be substantially increased by optimizing the hydrofracking process and, of course, drilling additional wells.

11.4 THE RATON BASIN (COLORADO AND NEW MEXICO)

The Raton Basin straddles southeastern Colorado and northeastern New Mexico, as shown in Fig. 11.6.

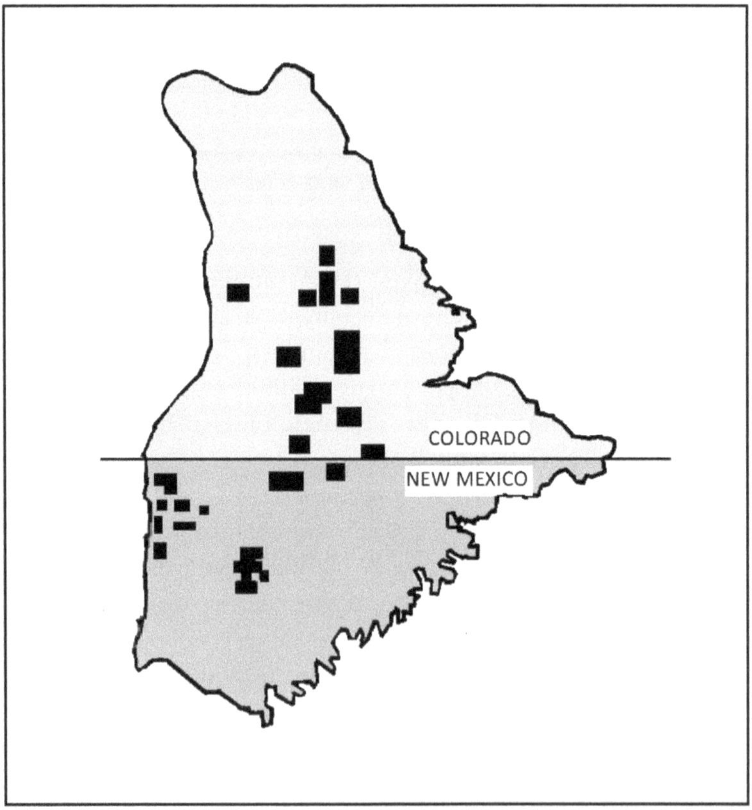

Figure 11.6 Major coalbed methane activity in the Raton Basin.

Table 11.3 Initial production and hydrofrack data for Raton Basin

Depth (ft)	Hydrofracked interval thickness (ft)	Estimated water (gal)	Sand (lbs)	Initial gas (MCFD)	Water (bpd)
1500–2000	20	400,000	445,000	70–100	40–80
2400	300	300,000	300,000	340	368
1800	300	400,000	474,000	117	20
1500[a]	280	500,000	532,000	161	115

[a]This may be a horizontal fracture.
Source: Adapted from Gas Research Institute. Methane from coal seams technology. 1993;11(1):52.

Several operators, such as Chevron, Meridian Oil, Pennzoil, and Western Oil are active in the area, producing 105 BCF/year. Reservoir properties vary considerably over the basin, but the north-south trending area in the center, as shown in Fig. 11.6, is the most promising.

Net coal thickness approaches 80 ft, but typically only 20 ft of the coal is stimulated for gas production to minimize water production. The rank of coal ranges from hvC near the outcrop to medium vol. bituminous in the deeper areas of the basin. The gas content is 400 ft^3/t at a depth of 1600–2000 ft. Total gas reserve is estimated at 11 TCF.

Vertical drilling with hydrofracking is the method used all over the basin for gas production. A typical job uses 100,000–800,000 gallons of water with 300,000–500,000 pounds of sand in two horizons. Initial gas production ranged from 100–350 MCFD with 100–600 bbl of water per day.

Data for some wells are presented in Table 11.3.

The below-average production is mainly due to the inability to hydrofrack the coal seams. The fracking fluid is apparently going into water aquifers producing large volumes of water. A better procedure would be to isolate coal seams that are at least 4–5 ft thick and hydrofrack them individually. The total coal thickness and high gas contents forecast a high gas production, in excess of 500 MCFD/well. Additional data on the reservoir is needed to optimize gas production.

REFERENCE

[1] Gas Research Institute. Methane from coal seams technology. 1993;11(1):52.

CHAPTER 12

Coalbed Methane Production From Deep Coal Reservoirs

Contents

Abstract

The largest reserves of coal and coalbed methane are in coal seams that are deeper than 3000 ft. There are four coal basins in the United States where CBM is being produced from deeper horizons the: (1) San Juan Basin, (2) Piceance Basin, (3) Green River Basin, and (4) Uinta Basin. Vertical drilling with hydrofracking is the most commonly used technique, although it is not the best technique. These four basins produce the majority (52–60%) of US CBM production. The San Juan Basin is over-pressurized and yields excellent gas production even if the completion technique is not optimal.

1. San Juan Basin: It is located at the western junction of Colorado and New Mexico. It is the highest producing basin at 650 BCF/year. It has a potential

Advanced Reservoir and Production Engineering for Coal Bed Methane.
DOI: http://dx.doi.org/10.1016/B978-0-12-803095-0.00012-0

reserve of 84 TCF in the Fruitland and Menefee formations. Because of over-pressurization, vertical wells do not always need hydrofracking. Cavitation (successive pressurization and depressurization) often yields equally good production. However, a much higher production of 5–6 MMCFD/well can be achieved by horizontal drilling and hydraulic stimulation as practiced in the Marcellus Shale of the northeastern United States. A typical design for hydrofracking the Fruitland coal seam is presented that can produce 1–2 MMCFD. A scheme of horizontal drilling is also presented that can lend itself to CO_2 flooding for secondary recovery and tertiary recovery by underground coal gasification.

2. Piceance Basin: This is the deepest basin in the United States, located in northwestern Colorado. It has a gas reserve of 84 TCF. Current annual production is only 5 BCF because the deep coal seams have very low permeability, so vertical wells do not produce much. The potential to produce larger quantities of gas by horizontal drilling and hydrofracking is great.
3. The (Greater) Green River Basin: It is located in northwestern Colorado and southwestern Wyoming. It is not well developed and produced only 20 BCF in 2015. The total gas reserve in the Greater Green River Basin is 83 TCF and there is great potential for increased gas production if horizontal drilling and hydrofracking is used for gas production.
4. The Uinta Basin: The basin is located mostly in Utah. It has a gas reserve of 9 TCF with an annual production of 42 BCF. While there is room to improve the vertical well completion technique, the recommended procedure for deep coal seams remains horizontal drilling from the surface with massive hydrofracking.

The largest reserves of coal and coalbed methane are in coal seams that are deeper than 3000 ft. These coal seams are beyond the economic limit for mining and are prime candidates for commercial gas production.

As shown in Chapter 10, Coalbed Methane Production From Shallow Coal Reservoirs, Fig. 10.1, there are four coal basins in the United States where coalbed methane is being produced from a depth horizon deeper than 3000 ft. They are

1. the San Juan Basin
2. the Piceance Basin
3. the Greater Green River Basin and
4. the Uinta Basin.

Together these basins produce nearly 52% of total US CBM production. Although these coal seams are the largest producers, very little is known about their reservoir properties. The permeability is, in general, low and gas content is high. The production technique is far from being optimized.

The prospect for producing very large quantities of gas from deep reservoirs is very good. The biggest impediment to high gas production is the very low permeability. Hydraulic stimulation, secondary recovery with CO_2 flooding, and in situ gasification of coal can enhance gas recovery significantly [1].

12.1 THE SAN JUAN BASIN (COLORADO AND NEW MEXICO)

This is the most productive CBM field in the United States. Drilling started in 1980 and in a mere 12 years, nearly 2000 wells were drilled, producing 450 BCF/year. Many large oil and gas companies are active in this basin. By 2012, nearly 4000 wells had been drilled, which produced 600 BCF/year.

Fig. 12.1 shows the geographical location of the basin in southwestern Colorado and northwestern New Mexico. The overall size of the basin is 21,000 square miles but the most productive area is only 7500 square miles, containing two major coal formations: (1) the Fruitland Formation, with a gas reserve of 50 TCF and (2) the Menefee Formation, with a gas reserve of 34 BCF.

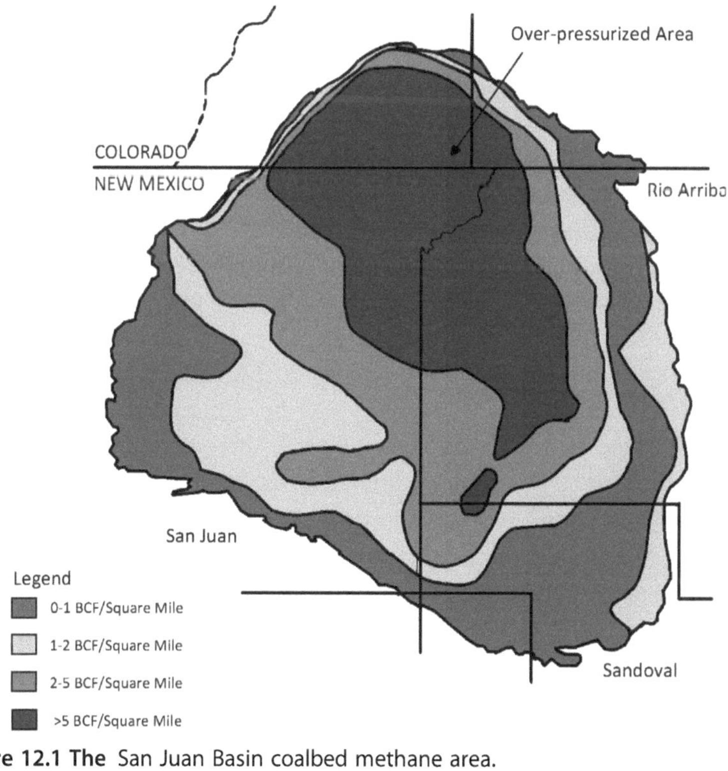

Figure 12.1 The San Juan Basin coalbed methane area.

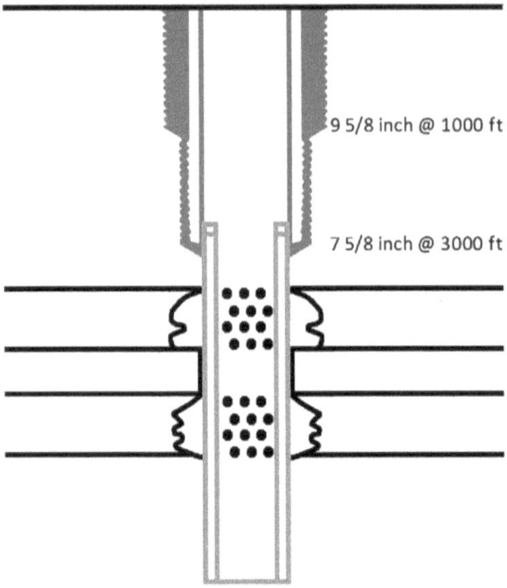

Figure 12.2 Open-hole cavitation completion.

During the 1980s, the standard completion practice was to complete the well by casing its entire depth and then hydraulically stimulate the well with linear gel, cross-linked gel, nitrogen foam, fresh water, or slick water with a number of additives. Later it was discovered that gas production can be dramatically increased by a different method, open-hole completion. Fig. 12.2 shows the completion process.

The coal seam is hydro-jetted to clean it and then successively pressurized and depressurized. This is known as the "cavitation process." Most likely the production improvement was achieved by minor fractures created in the coal when it was pressurized, but this was never confirmed. It is likely that even greater productivity can be achieved if the thick coal seams are drilled vertically and properly hydrofractured or drilled horizontally and then hydraulically stimulated. Both of these procedures to enhance gas production will be discussed.

12.1.1 Reservoir Characteristics

In spite of intensive drilling in the basin, there is very little information available on reservoir characteristics. The main gas-producing coal seam is the Fruitland Formation coal. It outcrops near the margin of the basin

Table 12.1 Characteristics of the Fruitland coal seam

Depth	3000–3500 ft
Thickness	40–60 ft
Rank of coal	Sub-bituminous
Gas content	400–600 ft^3/t
Gas composition	98–99% methane
	Less than 1% CO_2
Permeability	0.1–1.00 md
Porosity (helium)	3–5.7%
Compressive strength of coal	6000–8000 psi
Elastic modulus, E	$0.5–1.00 \times 10^6$ psi
Reservoir pressure	1–1.2 times the hydrostatic head

and contains 230 billion tons of coal in the entire basin to a depth of 4000 ft. Nearly 4000 wells have been drilled for gas production. The coal seam thickness averages 40 ft in the most productive area and reaches 60 ft in some areas. Reservoir characteristics of the Fruitland coal seam are shown in Table 12.1. This table is based on very limited available data [2].

Reservoir pressure is the most unusual feature of this basin. It is over-pressurized and hence highly productive in spite of poor completions. The coal seam also has a high diffusivity (short sorption time), indicating that 70–80% of the gas in the reserve can be recovered over a period of 15–20 years.

12.1.2 The Hydraulic Stimulation of a Vertical Well in the San Juan Basin

A good hydrofracking design tries to achieve

1. high fracture conductivity,
2. long propped length,
3. low frac fluid volumes, and
4. minimal chance of screen out.

A fracking design for a gas well is presented that will produce about 1 MMCFD for 15 years.

Depth	3500 feet (reservoir pressure = 1700 psi)
Net Pay	40 ft
Permeability	1 md
B.H.Temperature	100°F
Spacing	160 acres
r_e	1320 ft
r_w	0.25 ft
Average permeability of sand, 20/40 mesh	60,000 md

Assumptions:

Average created width	0.5 inch
Average propped width	$\leq$ 0.5 inch
Pumping rate	40 BPM = 3.73 ft^3/s = 224 ft^3/min
Fracture height	60 ft
Young's modulus, E	1×10^6 psi
Poisson ratio	0.35
Shear modulus, G	0.37×10^6 psi
Frac fluid	slick water with a viscosity of 1 cp
Leak-off coefficient, c	0.005 $ft/min^{0.5}$
c for design	0.0075 $ft/min^{0.5}$

Solution:

1. Find the frac length needed:

   Assuming a specific production for the Fruitland seam at 60 MCFD/100 ft, a total length of fracture of 1500 ft is needed. Hence half the fracture length is 750 ft. To create a propped length of 750 ft, a reasonable created length, *L*, would be 1000 ft (based on direct observation in other coal basins).

2. Select a frac fluid:

   Based on the success achieved with slick water in both coal and deeper shale formations, the frac fluid will be slick water with a viscosity of 1 cp.

3. Calculate the average frac width:

   Using the P-K model in Chapter 8, Hydraulic Fracking of Coal, Eq. (8.10):

$$\text{Average width} = \text{Wave} = 2.1\left(\frac{q\mu L}{E^1}\right)^{\frac{1}{4}}$$

$$= 2.1\left(\frac{2.08\times 10^{-5}\times 3.73\times 2000}{1.09\times 10^6}\right)^{\frac{1}{4}}$$

$$= 0.5 \text{ inch}$$

This is in an excellent agreement with the assumed width. This is the created average width. The propped width could be slightly less than 0.5 inch.

4. Calculate the fluid volume:

Area of the two wings of the fracture,

$$A = 2\times L\times H = 20,000\times 60 = 120,000 \text{ ft}^2.$$

The volume of frac fluid is given by Eq. (12.1):

$$2\ V^{0.5} = \left[\frac{3AC}{q^{0.5}}\right] + \left\{\left(\frac{3AC}{q^{0.5}}\right)^2 + 4AW\right\}^{\frac{1}{2}}$$

$$= \left[\frac{3\times 120,000\times 0.0075}{(224)^{0.5}}\right]$$

$$+ \left\{\left(\frac{3\times 120,000\times 0.0075}{(224)^{0.5}}\right)^2 + 4\times 120,000\times \frac{0.5}{12}\right\}$$

$$= 180 + \left\{(180)^2 + 20,000\right\}^{\frac{1}{2}}$$

$$= 180 + 229 = 409$$

$$\text{Hence } V = (204.5)^2 = 41,820 \text{ ft}^3 = 313,652 \text{ gallons}$$

$$= 7468 \text{ barrels} \tag{12.1}$$

5. Calculate the pumping time:

$$7408 \div 40 = 187 \text{ minutes.}$$

6. Calculate the fluid lost in the formation:

$$V_{FL} = A\ 3\ C\ T^{0.5} \text{ ft}^3$$

$$= 120,000\times 3\times 0.0075\ (187)^{0.5}$$

$$= 36,922 \text{ ft}^3$$

7. Select a pad volume (fluid pumped in the beginning without any sand, usually 40% V_{FL}):

$$= 0.4\ V_{FL}$$
$$= 0.4 \times 36,922\ \text{ft}^3 = 14,769\ \text{ft}^3 = 110,766\ \text{gallons} = 2687\ \text{barrels}$$

8. Select amounts of proppant and sand slurry:

Excellent fracture conductivity is obtained with 2 lbs of sand per square foot of the fracture area.

Hence total sand needed = 2 × 120,000 = 240,000 lbs.

$$\text{Volume of sand} = \frac{240,000}{22.1} = 10,860\ \text{gallons}$$

Hence the volume of water = 313,652 − 10,860 = 302,792 gallons.

9. Select volumes of water for ten stages from 1/4 to 3 ppg:

$$(313,052 - 110,766) \div 10 = 20,286\ \text{gallons}$$

The proppant schedule is shown in Table 12.2.

A well completed as discussed has the potential to produce 1–2 MMCFD in the over-pressurized area of the basin. Average production for vertical wells in the basin ranges from 200 to 2000 MCFD with an average of 600 MCFD/well [3].

Table 12.2 Frac plan for a vertical well in the San Juan Basin

Stage	Water volume (gallon)	ppg	Proppant lbs (1000 g)	Size of sand
1	110,000	0	pad	—
2	20,000	0.25	5	80–100 mesh
3	20,000	0.5	10	20–40 mesh
4	20,000	0.75	15	20–40 mesh
5	20,000	1.0	20	20–40 mesh
6	20,000	1.25	25	20–40 mesh
7	20,000	1.5	30	20–40 mesh
8	20,000	2.0	40	20–40 mesh
9	20,000	2.0	40	20–40 mesh
10	20,000	1.5	30	20–40 mesh
11	20,000	1.25	25	10–20 mesh
Total	310,000		240 k lb.	

Stage 11 is done with 10/20 sand to minimize loss of proppant during production.
Q = 40 bpm; $c = 0.0075\ \text{ft/min}^{0.5}$; W = 0.5 inch; L = 1000 ft.; H = 60 ft.

12.1.3 Gas Production From Hydrofracked Horizontal Wells

Application of horizontal wells with hydrofracking is a new innovation that revolutionized gas production from the Marcellus Shale in the northeastern United States. The 60–80-ft thick shale has a gas content of 75 ft^3/ton and permeability on the order of 0.01 md. In spite of these conditions, wells producing 5–6 MMCFD are quite common.

It is, therefore, logical to think that such wells in the Fruitland coal seam will produce at least this much gas per well. The coal seam has a much higher gas content (average 400 ft^3/t) and an order of magnitude higher permeability (0.1 md).

Fig. 12.3 shows a proposed drilling pattern. From a single location four sets of horizontal wells are drilled as shown. Each set of horizontal wells has three laterals in parallel, about 1000 ft apart. At a depth of 3000–4000 ft, the laterals can easily reach a length of 5000 ft. The horizontal laterals are cased and completed as discussed in Chapter 8, Hydraulic Fracking of Coal, and stimulated at 1000 ft intervals. A typical proppant schedule is also provided in Chapter 8, Hydraulic Fracking of Coal (Table 8.6). Only the central lateral is stimulated at 1000 ft intervals.

A well completed like this has a potential to produce 5–6 MMCFD, which is considerably higher than the production from a vertical well.

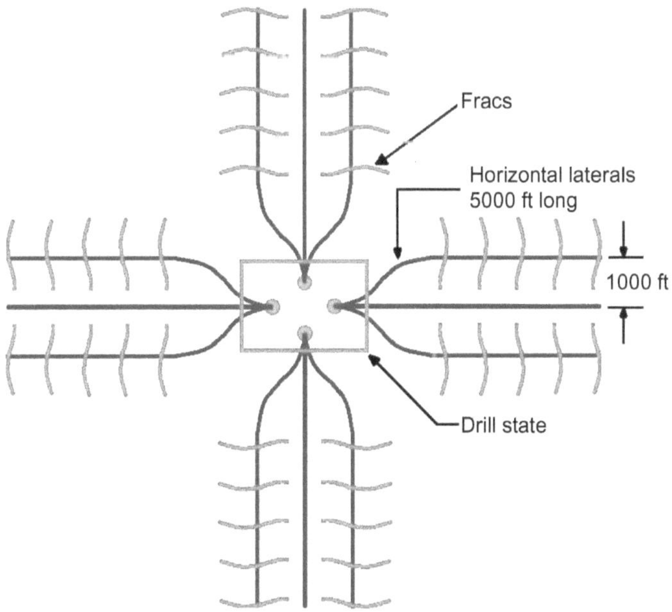

Figure 12.3 Horizontal well completion in deep coal seams.

The local economy will determine which technique is preferable, but horizontal wells have totally replaced vertical completions in the Marcellus Shale. Horizontal wells as proposed are much more expensive than a single vertical well.

12.2 THE PICEANCE BASIN

The Piceance Basin, in northwestern Colorado, is the deepest coalbed methane reserve in the United States, with the largest gas reserve, at 84 TCF. Because of the extreme depth (generally greater than 7000 ft), the coal seams are highly gassy and high in rank but have very low permeability (Fig. 12.4).

Extending over more than 6700 square miles, the Piceance Basin is comparable in area to the highly productive San Juan Basin. The lower production (5 BCF/year) is mainly due to low permeability of the CBM reservoir. More effective completion and stimulation, such as horizontal drilling with stimulation, can easily increase the gas production from this basin to 500 BCF/year.

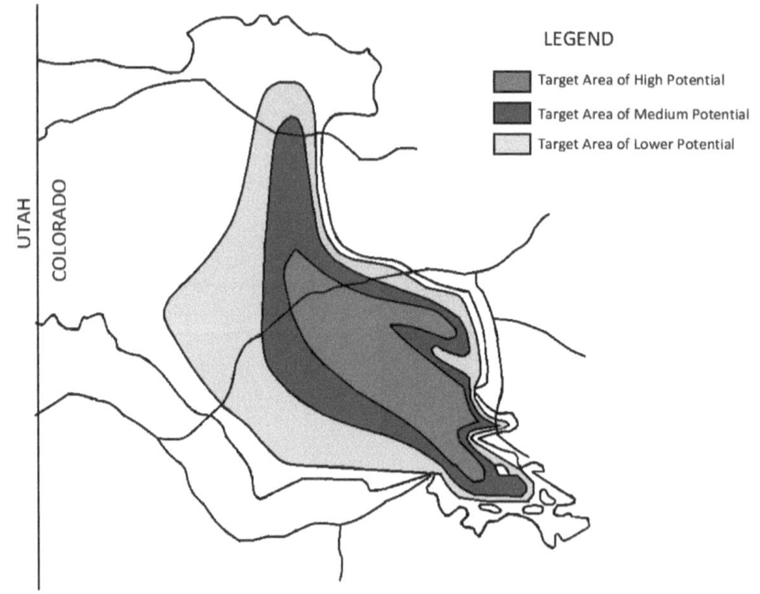

Figure 12.4 Piceance Basin coalbed methane prospects.

12.2.1 Coal Deposits of the Piceance Basin

The basin has been divided into eight coal fields, and their coal reserves are shown in Table 12.3. The rank of coal runs from anthracite to sub-bituminous depending on the depth.

However, coal reserve estimates using geophysical logs are much larger, at 382 billion tons [5]. Over 75% of these coal seams are deeper than 3000 ft. In each coal field, there are many coal seams with a thickness of 3–23 ft.

A typical sequence of coal seams is shown in Table 12.4 for the Williams Fork formation. This is a very good reserve with great potential for commercial gas production.

Table 12.3 Coal reserves of Piceance Basin [4]

Coal field	Coal reserve deeper than 6000 ft (million tons)	Thickness of coal (ft)
Book Cliffs	7200	7.8
Grand Mesa	8600	16.3
Somerset	8000	21.7
Crested Butte	1560	5.8
Carbondale	5200	27.4
Grand Hogback	3000	16.3
Danforth Hills	10,500	22.8
Lower White River	11,760	11.0
Total	55,820	

Table 12.4 Piceance Basin coal seams (arranged in descending order) (Williams Fork Formation)

Coal seam	Approximate thickness (ft)	Depth (ft)
Lion Canyon	8	4000
Montgomery	9	4000
Grinsted	9	4500
Comrike	22	4500
Agency	8	4500
Wesson	23	4500
Fairfield #2	10	4700
Fairfield	3–10	4700
Bloomfield	15	4800
Major	18	5000

12.2.2 Reservoir Properties of Piceance Basin Coal

In general, the deeper coal seams have permeabilities lower than 0.1 md. The coal seams are dry and do not produce much water. The gas content of coal north of the Colorado River ranges from 25 to 200 ft^3/t, and this area has a total reserve of 6–48 TCF. The gas content of coal south of the Colorado River ranges from 150–450 ft^3/t with a total reserve of 22–65 TCF. The reservoir pressure is estimated to be about 0.7 × hydrostatic heads.

12.2.3 Current Production Technology

Vertical wells, completed with hydrofracking, are the only method of gas production at present. Large vertical intervals (100–400 ft) were hydrofracked using 3000–4000 bbl of gelled water with 300,000–400,000 lbs of sand. Some typical production numbers are listed in Table 12.5.

In a few cases bigger frac jobs (10,000 bbl of gelled water with 1.3×10^6 lb of sand) have resulted in higher initial production, averaging 1 MMCFD, but the gas production declined sharply to an average of 400 MCFD in a year or so. The main limitation appears to be the low permeability. In spite of this, the basin is currently producing 100 BCF/year.

12.2.4 Proposed Improved Production Technology

It is obvious from prior experience that high production in this basin is limited to about 400 MCFD per well. The preferred production technology would be to drill the thickest coal seams horizontally as shown in Fig. 12.3. The central borehole in each set of three laterals should be hydrofracked, as proposed for the San Juan Basin. Depending on the thickness and gas content of the coal seam, gas production of 1–5 MMCFD can reasonably be expected. In view of the vast gas reserve of this basin, horizontal drilling and hydrofracking is highly recommended.

Table 12.5 Gas production from Piceance Basin wells

Coal field	Peak production (MCFD)	Average production (MCFD)
Grand Valley	600	200
White River	400	300
Parachute Field	250	150
South Shale Ridge	<100	<100

12.3 THE GREATER GREEN RIVER BASIN

The Greater Green River Basin is structurally complex and covers 21,000 square miles in southwestern Wyoming and northwestern Colorado. The rank of coal ranges from sub-bituminous with modest gas contents to medium volatile coal with gas contents of 400 ft^3/t. The basin is not well developed and produced just 16 BCF/year in 2015.

The Greater Green River Basin can be divided into four distinct basins—Sand Wash and Washakie, Great Divide, Rock Springs, and Green River (Fig. 12.5).

The Mesa Verde Group (Upper Cretaceous), Fort Union Formation (Paleocene), and Wasatch Formation (Eocene) are the primary coal-bearing intervals, with a net coal thickness of up to 150 feet. Coal deposits are 6000 ft deep on the shallow side (basin margins) but the depth increases to 12,000 ft near the center of the basin.

The geologically distinct Hanna Basin to the east is also included in the Green River coal region.

Table 12.6 shows the estimated coal resources [6] to a depth of 3000–6000 ft.

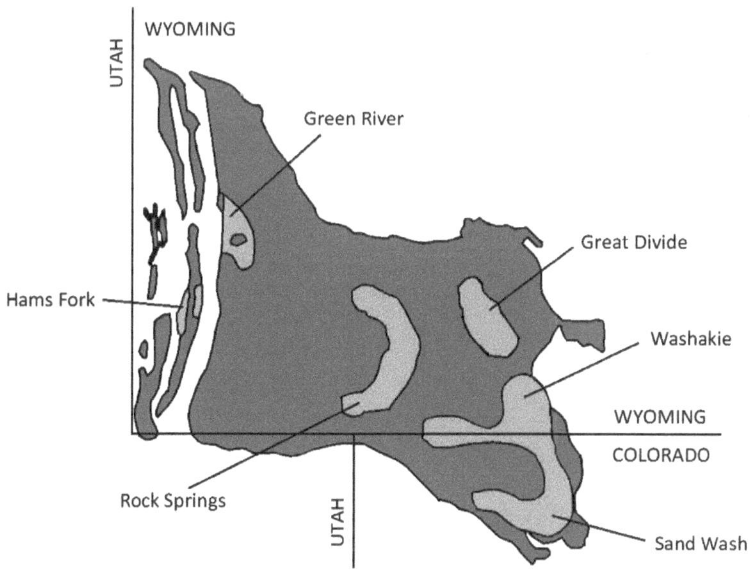

Figure 12.5 The Greater Green River methane prospects.

Table 12.6 Coal resources of the Greater Green River and Hams Fork Basins

Coal basin	Coal fields	Estimated reserve (billion ton)[a]	Gas reserve (TCF)	Average thickness (ft)
Greater Green River	Rock Springs	14	29	N/A
	Great Divide	4		N/A
	Washakie/Sand Wash	57		4–35
	Green River	3		5–90
Hams Fork		5	1	4–35
Total		83	30	—

[a]These are very conservative estimates.

12.3.1 Gas Production From the Sand Wash Basin

In early 1990, Fuelco drilled nine wells in Carbon County, Wyoming, to depths of 1700–2500 ft. Selected coal formations were hydrofracked with 1000–2500 bbl of water with 60,000–500,000 lbs of sand. Shallow wells did not produce much gas (6–60 MCFD/well) because the gas content to a depth of 2000 ft is very low (10–50 ft^3/t). The shallowest well (about 1000 ft deep), however had a horizontal fracture and was the most productive.

Cockrell Oil Corporation also drilled a cluster of 16 wells at 160-acre spacing. Because of poor completion none of these produced any gas. The hydraulic fracture job was too small to succeed (900 bbl of water with 70,000 lbs of sand).

12.3.2 The Great Divide Basin

Triton Oil and Gas drilled four wells but the results were disappointing. The depth of the wells was 3200–4200 feet. Average gas content was high at 270 ft^3/t. The production averaged 70–80 MCFD with 200 bpd of saline water. The TDS concentration was 10,000–29,000 ppm, which required treatment of the water in evaporation ponds. The main reason for poor gas production is the low permeability of the coal seams. Vertical wells at a depth of 3000 ft generally do not produce well, unless the coal seam is over-pressurized.

12.3.3 The Green River Basin

Buttonwood Petroleum of Oklahoma drilled a number of wells to depths of 5000–6000 ft. The wells were just hydro-jetted across five to six coal seams with a total thickness of 20 ft. No gas production was realized because low-permeability coal at this depth is not likely to produce any gas.

12.3.4 The Washakie Basin

The wells were completed at 2500–3174 ft depth with hydraulic stimulation using 1000–4000 bbl of water and 100,000–650,000 lbs of sand. The best gas production was about 500 MCFD. This is one of the best results from this basin.

Similar efforts in the Hanna Basin, however, met with failure. The coal seam depth (4000–6000 ft) precluded success because of the poor permeability.

12.3.5 Proposed Production Technology

It is clear from the above data on many coal fields of this basin that the only successful technique is to drill long horizontal boreholes in the thicker coal seams and hydraulically stimulate them as proposed for the San Juan Basin. Because of lower gas content and greater depth, the gas production per well is not likely to be as high as for the San Juan Basin but it can be commercially viable at gas prices above $5/MCF.

12.4 THE UINTA BASIN

The Uinta Basin covers an area of 14,450 square miles mostly in the northeastern part of Utah and a small area of northwestern Colorado (Fig. 12.6).

The basin is stratigraphically contiguous with the Piceance Basin but separated structurally from it by the Douglas Creek arch. Coal seams of more than 20 feet net thickness, with high rank, occur at a depth of 2000–4500 ft. The Utah Geological Survey estimated the coal reserves of the basin at 30 billion tons to a depth of 9000 ft. The average gas content for coal seams at depths of 1000–3000 ft is 300 ft^3/t. The gas reserve is, therefore, about 9 TCF. The four major coal fields in the basin are Emery, Wasatch Plateau, Sego, and Book Cliffs (lying between Wasatch and Sego). The total gas production from the entire basin was 42 BCF/year in 2015. Most of the drilling has been done in the Wasatch and Book Cliffs coal fields because of their proximity to gas pipelines.

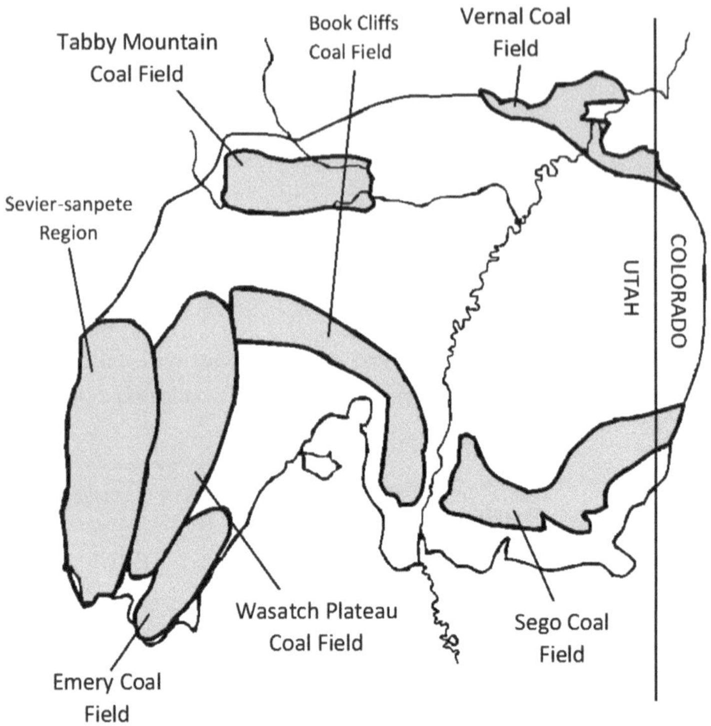

Figure 12.6 The Uinta Basin coal fields.

12.4.1 The Book Cliffs Coal Field

PG&E was the first to drill in this area. The wells were drilled at 160-acre spacing to a depth of 4000–4500 ft. The wells were hydrofracked with 80,000–140,000 lbs of sand using water as the carrying fluid. Average gas production was 120 MCFD with 300 bpd of water. Gas composition is generally 90% methane with 10% carbon dioxide. The water is brackish and needs special treatment before discharge.

Another operator, River Gas Corporation, drilled several wells in this coal field but to a shallow depth of 1000–2000 ft. There was 20–50 ft of coal in a 100 ft interval that was hydrofracked. Most of the hydrofracking was done with cross-linked borate gel with high viscosity, which can carry heavy sand loading (8–9 ppg). The average amount of sand used was 250,000 lb/well. The average gas production was 45 MCFD with 105 bpd of water. The fractures were most likely horizontal, explaining the poor gas production.

12.4.2 The Wasatch Coal Field

Cockrell Oil drilled two very deep (about 7500 ft) wells in this field. The wells were not hydrofracked and therefore did not produce any gas. Anadarko Petroleum Corporation also drilled some wells at depths of 4000–7000 feet. Gas production data is not available but production is likely to be poor because of low permeability.

12.4.3 Proposed Production Technology

Similar to other deep coal basins, it is safe to conclude that vertical drilling and hydrofracking will not yield commercial quantities of gas in this coal field. The preferred technique would be to drill thick (40–60 ft) coal seams horizontally and hydrofrack the laterals as discussed before.

12.5 SECONDARY RECOVERY BY CARBON DIOXIDE FLOODING

When the diffusivity of the coal seam is low, the ultimate recovery of in situ coalbed methane may be as low as 50%. The recovery can be enhanced by displacing methane by a gas with smaller molecular diameter, such as carbon dioxide. The process is called secondary recovery of coalbed methane.

It is a well-known fact that coal has a great affinity for CO_2, and CO_2 can displace methane stored on the micropores of the coal matrix. This property of coal has been used to enhance methane production from deep coal seams and simultaneously sequester CO_2 in the coal. Laboratory experiments to displace methane from coal using nitrogen, helium, and carbon dioxide show that CO_2 is the most effective flooding agent. The storage capacity of coal for CO_2 is at least twice its capacity for methane [7,8].

Fig. 12.3 shows an ideal layout for CO_2 flooding for enhanced methane recovery. When gas production begins to decline and total recovery is approaching 40–50%, CO_2 injection in the middle lateral (which was hydraulically fractured before) can begin. The optimum parameters of CO_2 injection have not yet been established, but many research projects are in progress to determine the following:

1. Optimum storage capacity for CO_2 in various coal seams.
2. Rate of injection and the travel velocity of CO_2 in coal. It is typically very slow.

3. Optimum injection pressure: generally the reservoir fracturing pressure should not be exceeded.
4. Economics of CO_2 flooding.

CO_2 sequestration in coal seams is encouraged by many countries because CO_2 is considered a greenhouse gas. In some countries, there are financial incentives for sequestering CO_2. Coal is a source of CO_2 in two ways: (1) from power plants that burn coal and (2) the CO_2 in coalbed methane. In both cases CO_2 can be stripped using molecular sieves [9].

The combined revenues from enhanced methane production and CO_2 sequestration can make many deep coal seam ventures financially viable.

12.6 TERTIARY RECOVERY OF COALBED METHANE/UNDERGROUND COAL GASIFICATION

When CO_2 injection ceases to enhance coalbed methane production, the last resort is to somehow heat the coal seam. As discussed before, heating greatly increases the diffusivity of coal, resulting in increased gas recovery. This is called "tertiary recovery" of CBM.

Fig. 12.7 shows a hypothetical schematic for underground coal gasification. Heating of the central lateral (horizontal borehole) can be done with steam or radio frequency energy, but the most synergistic approach is to ignite the coal with a limited supply of air/oxygen combustible gas

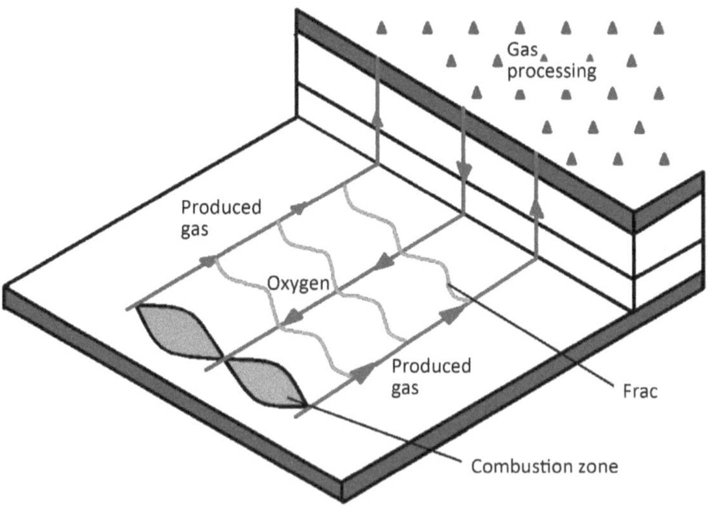

Figure 12.7 A schematic for underground coal gasification.

mixture. The underground coal gasification process is, in fact, already a known technique, but it has not been done in conjunction with methane recovery. In this scheme, the outer laterals are used for gas production while the central lateral is burning coal. While the methane thus produced has a relatively high calorific value, the underground coal gasification process produces low-calorific-value gas (generally less than 500 BTU/ft^3)

The composition of the gas produced will depend on coal composition and process parameters, such as operating pressure, outlet temperature, and gas flow. These parameters are constantly monitored and frequently adjusted to optimize the process.

The data on global coal deposits discussed in Chapter 1, Global Reserves of Coal Bed Methane and Prominent Coal Basins, indicates that only 3–6% of the total reserves are minable. The rest must be used to produce CBM and ultimately subjected to underground coal gasification. The low-BTU gas produced by underground coal gasification can be utilized in a variety of ways, including

1. as boiler fuel to produce steam for electric power generation,
2. as feed for chemical plants,
3. as feed for the Fischer-Tropsch process for producing liquid fuels such as diesel or aviation fuel, and
4. as a clean gaseous fuel after the raw gas is processed to produce pipeline-quality gas (calorific value $\geqq$ 960 BTU/ft^3).

Successful application of CO_2 flooding followed by underground coal gasification can unleash a vast resource, namely the natural gas residing in 17 to 30 T tons of coal in the global coal deposits.

REFERENCES

[1] Thakur PC. Coal-bed methane production 3rd edition. Vol. 2. SME Handbook of Mining Engineering; 2011. p. 1121–31.
[2] Jones AH, et al. Methane production characteristics for a deeply buried coal bed reservoir in the San Juan Basin, SPE/DOE/GRI Unconventional Gas Recovery Symposium, Pittsburgh, PA; 1984, p. 417–933.
[3] Gas Research Institute. *Quarterly review of methane from coal seam technology.* 1993;4(1): 1–110.
[4] Hornbaker AL, et al. Summary of coal resources in Colorado, Vol. 9. Colorado Geological Survey, Special Publication; 1976.
[5] Collins BA. Coal deposits of Carbondale,Grand Hogback, and Southern Danforth Hills coal fields, Colorado School of Mine, Quarterly, 71, 1; 1977. p. 138.
[6] Mroz TH, editor. Methane Recovery from Coal Beds: A Potential Energy Source. US Department of Energy; 1983. p. 458.

[7] Collins RC, et al. Modeling CO_2 sequestration in abandoned mines: Proceedings of AICHE Spring meeting, New Orleans, LA; 2002.
[8] Thakur PC, et al. Coal bed methane production from deep coal seams, Proceedings of the AICHE Spring meeting, New Orleans, LA; 2002.
[9] Mitariten MF, One-step removal of nitrogen and carbon dioxide from coal seam gas with the molecular gate system,The 2nd Annual CBM and CMM Conference, Denver, Colorado; 2001.

APPENDIX 1

Evaluation of Error Function

In mathematics, the error function (also called Gauss Error Function) is a special function of sigmoid shape as shown in Fig. A.1 that is related to the probability density of a standardized random variable.

It is defined as,

$$f(x) = \frac{1}{\sqrt{2\pi}} e^{-\frac{1}{2}x^2} \tag{A.1}$$

By definition, error function of x, or

$$\mathrm{erf}(x) = \frac{2}{\pi} \int_0^{x} e^{-x^2} dx \tag{A.2}$$

Let as substitute for $x = \left(\frac{x}{\sqrt{2}}\right)$ and $dx = \left(\frac{1}{\sqrt{2}}\right) dx$ in Eq. (A.2).

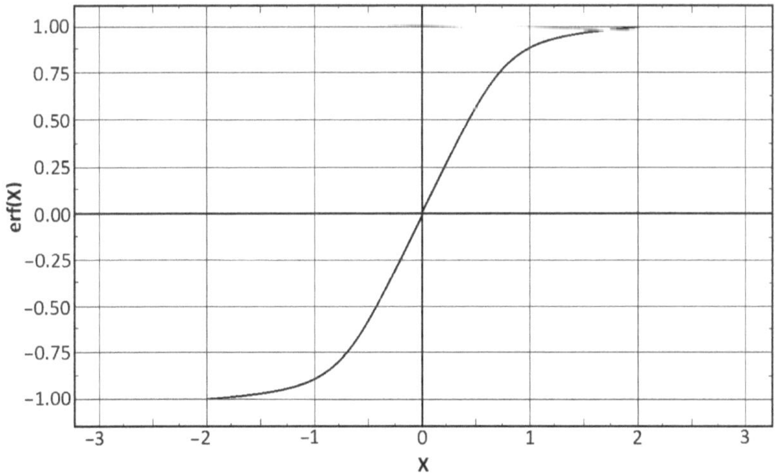

Figure A.1 Appendix A, Plot of Error Function.

Eq. (A.2) becomes

$$\operatorname{erf}\left(\frac{x}{\sqrt{2}}\right) = \frac{2}{\pi}\int_0^x e^{-\frac{x^2}{2}} \cdot \frac{1}{\sqrt{2}}\,dx$$

$$\text{or } \operatorname{erf}\left(\frac{x}{\sqrt{2}}\right) = 2\frac{1}{\sqrt{2\pi}}\int_0^x e^{-\frac{x^2}{2}}\,dx$$

$$\text{or } \operatorname{erf}\left(\frac{x}{\sqrt{2}}\right) = 2\mathrm{F}(x)$$

Where F(x) is cumulative distribution of the probability density function in Eq. (A.1).

Table A.1 shows the value of F(x) for x over a range of 0 to 4 at salient intervals. For intermediate values reference can be made to standard tables.

For values of x in between the numbers listed, a linear estimation can be done but for accurate results standard tables should be used.

To calculate erf(x) for a given (x) an example is shown. To evaluate erf (2.0) one proceeds as follows:

Since

$$\frac{x}{\sqrt{2}} = 2.0; \quad x = 2.828 = 2.83$$

Table A.1 Cumulative probability as a function of (x)

(x)	F(x)
0.00	0.5000
0.10	0.5199
0.20	0.5793
0.30	0.6179
0.40	0.6554
0.50	0.6915
0.60	0.7257
0.70	0.7580
0.80	0.7881
0.90	0.8159
1.00	0.8413
2.00	0.9773
3.00	0.9987
4.00	1.0000

From the tables $F(x) = 0.9977$. Subtracting 0.5 from the value of $F(x)$, one gets 0.4977. Thus $\text{erf}(2.0) = 2 \times 0.4977 = 0.9954$.

The complementary error function is denoted erfc and defined as:

$$\begin{aligned} \text{erfc}(x) &= 1 - \text{erf}(x) \\ &= \frac{2}{\pi}\int_{x}^{-\infty} e^{-x^2}\,dx \end{aligned} \tag{A.3}$$

The imaginary error function, denoted by erf*i* is defined as:

$$\text{erf}(x) = -i\,\text{erf}(ix)$$

The complex error function, denoted $w(x)$ and also known as Faddeeva function is defined as:

$$w(x) = e^{-x^2}\text{erfc}(-ix) = e^{-x^2}[1 + i\text{erf}\,i(x)]$$

P. Thakur
ESMS LLC (Expert Solutions to Mine Safety),
Morgantown, WV, United States

APPENDIX 2

Solutions to Unsteady State Flow Equations Constant Production Rate Infinite Radial System

Dimensionless time t_D	Pressure changes P_t
0.00	0.0000
0.001	0.0352
0.01	0.1081
0.1	0.3144
1.0	0.8019
2.0	1.0195
3.0	1.1665
5.0	1.3625
10.0	1.6509
20.0	1.9601
100.0	2.7233
500.0	3.5164
1000.0	3.8584

For $t_D > 1000$:

$$P_t = \frac{1}{2}(1n\ t_D + 0.80907)$$

Adapted from Katz DL, et al. Handbook of natural gas engineering. New York, NY: McGraw-Hill Book Company; 1958.

GLOSSARY OF TERMS

GLOSSARY OF TERMS CHAPTER 1

No Symbols.

GLOSSARY OF TERMS CHAPTER 2

Symbol	Description	Units
STP	Standard temperature and pressure	32°F 14.7 psi
Q	Cumulative volume of gas desorbed	ft^3
A	A characteristic of coal	ft^3/min (day)
t	Time	min or days
n	A characteristic constant	dimensionless
B	A measure of lost gas	ft^3
V	Volume of gas contained at pressure, P	ft^3/t
V_m	Maximum sorption capacity of coal	ft^3/t
P	Pressure	psi
b	A Langmuir constant	psi^{-1}
P_L	Langmuir pressure where V = Vm/2	psi
m	A characteristic constant of coal	dimensionless
k	A characteristic constant of coal	dimensionless
$V\phi$	Volume of gas in pores	ft^3/t
ϕ	Porosity	fraction
P_O	Atmospheric pressure	psi
T	Temperature	Kelvin
V_C	Volume of coal	ft^3/t
W	Moisture content	%
A	Ash content	%
G	Gas reserve	ft^3
A	Area in Cg (2.11)	acres
H	Thickness of coal	ft
C_g	Gas content of coal	ft^3/ft^3
MMCFD	Million cubic feet per day	–
BCF	Billion cubic feet	–
c_p	Specific heat of a gas	calories/ft^3

GLOSSARY OF TERMS CHAPTER 3

Symbol	Description	Units
ϕ	Porosity	%
V_P	Connected pore volume	ft^3
V_b	Bulk volume	ft^3
V_m	Volume of solid matrix	ft^3
u	Average fluid velocity	cm/s
A	Cross-sectional area	cm^2
k	Permeability	darcy
μ	Viscosity of fluid	centipoise
Q	Rate of fluid flow	cm^3/s
q	Q, normalized for average pressure	cm^3/s
P_1, P_2	Upstream and downstream pressure	psi
P_b	Average pressure	psi
b, (b_1, b_2)	Fracture width in coal	mm
A (Chapter 3)	Side of a fractureless cube of coal	mm
P_C	Closure pressure	psi
P_S	Bottom hole pressure in drawdown test	psi
P_f	Wellhead pressure	psi
Z	Compressibility of gas	fraction
T	Absolute temperature	Rankine
H	Thickness of coal	ft
μ_p	Gas viscosity	centipoise

Refer to Chapter 5 for symbols in drawdown and build-up tests

K_o	Permeability of coal at 100-ft depth	md
D	Depth	ft
$\sigma = (\sigma_H - \sigma_O)$	(Major horizontal stress − pore pressure)	psi
T	Absolute temperature	Rankine
E_V	Volumetric strain due to stress change	%
E_P	Volumetric strain due to desorption shrinkage	%
R	Gas content	–

GLOSSARY OF TERMS CHAPTER 4

Symbol	Description	Units
c	Number of molecules diffusing	g-mole
D	Diffusivity coefficient	cm/s^2
$\frac{dc}{dt}$	Rate of diffusion	g-
	mole/s	*a*
The	radius of a coal particle	cm
T	Time	s

Symbol	Description	Units
M_t	Amount of gas desorbed in time, t	cm^3
M_∞	Langmuir volume	cm^3
τ	Sorption time (time taken to desorb 63% of total gas)	days
S	Spacing between major cleats	cm
$Deff_A$	Effective average diffusion coefficient for A into a complex mixture	cm/s^2
D_{AB}	The order of the subscripts indicates diffusion of A into system AB	cm/s^2

GLOSSARY OF TERMS CHAPTER 5

Symbol	Description	Units
σ_O	Reservoir (pore) pressure	psi
σV	Vertical stress	psi
σH	Major horizontal stress	psi
σh	Minor horizontal stress	psi
D	Depth of the coal seam	ft
V_P	Velocity of pressure waves	μs/ft
V_S	Velocity of shear waves	μs/ft
ρ	Density of coal	lbs/ft^3
υ	Poisson ratio	dimensionless
E	Young's modulus of elasticity	psi
K	Bulk modulus of elasticity	psi
G	Shear modulus of elasticity	psi
μs	Velocity of sound in micro seconds to travel one foot	μs

GLOSSARY OF TERMS CHAPTER 6

Symbol	Description	Units
P, P_w, P_e	Pressure, pressure at well bore, pressure at boundary	psi
r	Radius	ft
μ	Viscocity	cp
φ_c	Pseudo-porosity for coal	fraction
k	Permeability	md
$\overline{p}$	Average of pressure at well and at infinite radius	psi
t	Time	s
erfc	Complementary error function	–
t_D	Dimensionless time	–
m	Flow rate	dimensionless
p_t	Pressure changes	dimensionless
z	Compressibility factor	fraction
Q	Gas production rate	MCFD

Symbol	Description	Units
h	Height of coal seam	ft
T	Temperature	Rankine
Q_T	Total cumulative gas production at time t	MCFD
Q_t	Gas influx	function of t_D
t_f	The length of time the well was flowing at a constant dimensionless rate, m	—
m_1	Dimensionless flow rate	—
q	Gas production in ft^3/day at STP	CFD
q_i	Initial production	MCFD
q_t	Production at time t	MCFD
d	Rate of decline	—

GLOSSARY OF TERMS CHAPTER 7

Symbol	Description	Units
H	Head loss	ft
λ	Frictional coefficient	dimensionless
l	Length of pipe	ft
v	Velocity of fluid	ft/s
g	Gravitational constant	ft/s^2
d	Pipe diameter	ft
ρ	Density of fluid	lb/ft^3
μ	Viscosity of fluid	cp
e	Pipe roughness	dimensionless
C	Constant of proportionality	—
Q	Rate of fluid flow	ft^3/hr or ft^3/day
R	Reynolds number	—
P	Pressure loss	ft
G	Specific gravity of a gas	—
T	Temperature	Rankine
Z	Compressibility factor	—
V_L	Minimum velocity of transport	ft/s
F_L	Proportionality constant	—
S	Specific gravity of a solid	—
S_L	Specific gravity of a liquid	—
C_V	Volumetric concentration	%
C_W	Weight concentration	%
λ_W	Frictional coefficient for water	—
λ_S	Frictional coefficient for slurry	—
η	Efficiency of an electric motor	—
K	Ratio of C_p/C_v for a gas	—

Symbol	Description	Units
P_1, P_2	Suction and discharge pressure	psi
E	Experimental constant	–
S	$\frac{0.0375G\ X}{T\ Z}$	–
W	Work done	ft-lb/lb
K	C_P/C_V	ratio
d_{eff}	$d_2 - d_1$ (well diameter − pipe diameter)	ft
d_h	Hydraulic diameter	ft
A_C	Cross-sectional area of a non-circular pipe	ft^2

GLOSSARY OF TERMS CHAPTER 8

Symbol	Description	Units
V	Total fluid volume	ft^3 or gallons
c	Leak-off coefficient	$ft/\sqrt{min}$
V_F	Fracture volume	ft^3
V_L	Fluid loss in coal	ft^3
Q	Rate of pumping	ft^3/min
T	Time	min or s
L	Length of fracture	ft
W	Width of fracture	ft
H (H_P)	Height (wetted height) of fracture	ft
η	Efficiency of fracking	–
υ	Poisson ratio	–
P	Fracture pressure	psi
σ	Pore pressure	psi
$E^1 = \left(\frac{E}{1-\upsilon^2}\right)$	Plain strain modulus	psi
W_{MAX}	Maximum theoretical width of fracture	ft
P_b	Bottom hole pressure	psi
P_c	Closure pressure	psi
μ	Viscosity $\left(1\ cp = 2.08 \times 10^{-5}\ \frac{lb\text{-}sec}{ft^2}\right)$	$\frac{lbs}{ft^2}$
F.G.	Fracture gradient	psi/ft
t_o	Total time of pumping	min
t_{si}	Time interval when pressure reading was taken after t_o	min

GLOSSARY OF TERMS CHAPTER 9

Symbol	Description	Units
γ	Thrust needed to drill	lbs
X_1	Length of borehole	ft

Symbol	Description	Units
X_2	Pressure differential across the drill motor	psi
X_3	Rate of drilling	ft/min
X_4	Torque	lb-inch
$\Delta\theta$	Angle build in 10 ft	degrees
T	Thrust	lb

GLOSSARY OF TERMS CHAPTER 10

No Symbols.

GLOSSARY OF TERMS CHAPTER 11

No Symbols.

GLOSSARY OF TERMS CHAPTER 12

Symbol	Description	Units
r_e	Radius of the reservoir	ft
r_w	Radius of the well	ft
μ	Viscosity of fluid	cp
q	Rate of pumping	ft^3/s
L	Length of fracture	ft
H	Height of fracture	ft
A	Wetted area of frac (L x H)	ft^2
W	Width of fracture	ft
c	Leak-off coefficient	$ft/\sqrt{min}$
V	Total volume of fluid	ft^3
V_{FL}	Volume of fluid lost	ft^3
BTU	British thermal unit	

INDEX

Note: Page numbers followed by "*f*" and "*t*" refer to figures and tables, respectively.

G

Kolibri 360 GmbH
Ferdinand-Jühlke-Straße 7
99095 Erfurt, Deutschland
produktsicherheit@kolibri360.de